Communications in Computer and Information Science 2615

Series Editors

Gang Li ⓘ, *School of Information Technology, Deakin University, Burwood, VIC, Australia*
Joaquim Filipe ⓘ, *Polytechnic Institute of Setúbal, Setúbal, Portugal*
Zhiwei Xu, *Chinese Academy of Sciences, Beijing, China*

Rationale

The CCIS series is devoted to the publication of proceedings of computer science conferences. Its aim is to efficiently disseminate original research results in informatics in printed and electronic form. While the focus is on publication of peer-reviewed full papers presenting mature work, inclusion of reviewed short papers reporting on work in progress is welcome, too. Besides globally relevant meetings with internationally representative program committees guaranteeing a strict peer-reviewing and paper selection process, conferences run by societies or of high regional or national relevance are also considered for publication.

Topics

The topical scope of CCIS spans the entire spectrum of informatics ranging from foundational topics in the theory of computing to information and communications science and technology and a broad variety of interdisciplinary application fields.

Information for Volume Editors and Authors

Publication in CCIS is free of charge. No royalties are paid, however, we offer registered conference participants temporary free access to the online version of the conference proceedings on SpringerLink (http://link.springer.com) by means of an http referrer from the conference website and/or a number of complimentary printed copies, as specified in the official acceptance email of the event.

CCIS proceedings can be published in time for distribution at conferences or as post-proceedings, and delivered in the form of printed books and/or electronically as USBs and/or e-content licenses for accessing proceedings at SpringerLink. Furthermore, CCIS proceedings are included in the CCIS electronic book series hosted in the SpringerLink digital library at http://link.springer.com/bookseries/7899. Conferences publishing in CCIS are allowed to use Online Conference Service (OCS) for managing the whole proceedings lifecycle (from submission and reviewing to preparing for publication) free of charge.

Publication process

The language of publication is exclusively English. Authors publishing in CCIS have to sign the Springer CCIS copyright transfer form, however, they are free to use their material published in CCIS for substantially changed, more elaborate subsequent publications elsewhere. For the preparation of the camera-ready papers/files, authors have to strictly adhere to the Springer CCIS Authors' Instructions and are strongly encouraged to use the CCIS LaTeX style files or templates.

Abstracting/Indexing

CCIS is abstracted/indexed in DBLP, Google Scholar, EI-Compendex, Mathematical Reviews, SCImago, Scopus. CCIS volumes are also submitted for the inclusion in ISI Proceedings.

How to start

To start the evaluation of your proposal for inclusion in the CCIS series, please send an e-mail to ccis@springer.com.

Lukas Fischer · Ulrich Göhner ·
Sebnem Gül-Ficici · Dirk Jacob ·
Gabriele Kotsis · A Min Tjoa · Ismail Khalil
Editors

Database and Expert Systems Applications - DEXA 2025 Workshops

AISys and AI4IP
Bangkok, Thailand, August 25–27, 2025
Proceedings

Editors
Lukas Fischer ⓘ
Software Competence Center Hagenberg
Hagenberg, Austria

Sebnem Gül-Ficici ⓘ
University of Applied Sciences Kempten
Kempten, Germany

Gabriele Kotsis ⓘ
Johannes Kepler University Linz
Linz, Austria

Ismail Khalil ⓘ
Johannes Kepler University Linz
Linz, Austria

Ulrich Göhner ⓘ
University of Applied Sciences Kempten
Kempten, Germany

Dirk Jacob ⓘ
University of Applied Sciences Kempten
Kempten, Germany

A Min Tjoa ⓘ
Vienna University of Technology
Vienna, Austria

ISSN 1865-0929　　　　　　ISSN 1865-0937　(electronic)
Communications in Computer and Information Science
ISBN 978-3-032-02002-4　　　ISBN 978-3-032-02003-1　(eBook)
https://doi.org/10.1007/978-3-032-02003-1

© The Editor(s) (if applicable) and The Author(s), under exclusive license
to Springer Nature Switzerland AG 2026

This work is subject to copyright. All rights are solely and exclusively licensed by the Publisher, whether the whole or part of the material is concerned, specifically the rights of translation, reprinting, reuse of illustrations, recitation, broadcasting, reproduction on microfilms or in any other physical way, and transmission or information storage and retrieval, electronic adaptation, computer software, or by similar or dissimilar methodology now known or hereafter developed.
The use of general descriptive names, registered names, trademarks, service marks, etc. in this publication does not imply, even in the absence of a specific statement, that such names are exempt from the relevant protective laws and regulations and therefore free for general use.
The publisher, the authors and the editors are safe to assume that the advice and information in this book are believed to be true and accurate at the date of publication. Neither the publisher nor the authors or the editors give a warranty, expressed or implied, with respect to the material contained herein or for any errors or omissions that may have been made. The publisher remains neutral with regard to jurisdictional claims in published maps and institutional affiliations.

This Springer imprint is published by the registered company Springer Nature Switzerland AG
The registered company address is: Gewerbestrasse 11, 6330 Cham, Switzerland

If disposing of this product, please recycle the paper.

Preface

Welcome to the Proceedings of the DEXA 2025 Workshops. This year, we hosted two workshops: the 5th International Workshop on AI System Engineering: Math, Modelling and Software (AISys 2025) and the 1st International Workshop on Optimisation of Industrial Production with AI Algorithms (AI4IP). These events took place from August 25–27, 2025 in Bangkok, Thailand.

This compilation of papers and presentations represents a convergence of cutting-edge research, interdisciplinary collaboration, industrial application of algorithms and methods, and innovative solutions at the forefront of science and technology.

AISys 2025 focused on the foundational aspects of trustworthy AI systems. The workshop brought together scientists from artificial intelligence, systems engineering, applied mathematics, and software architecture to explore the foundational and practical challenges of building robust, scalable, and trustworthy AI systems. This workshop also addressed the ethical dimensions of AI systems, with particular attention to compliance challenges arising from emerging regulatory frameworks such as the European AI Act. The contributions presented in this workshop explore key issues including the trustworthiness of AI systems, risk mitigation strategies for large language model (LLM) outputs, and the algorithmic and architectural challenges encountered in the development and deployment of complex AI systems. Together, these papers provide valuable insights into the intersection of ethical principles, regulatory demands, and technical implementation.

AI4IP 2025 was initiated by a group of scientists working in the very fast-growing field of AI algorithms applied to manufacturing processes. Today's production lines and assembly workflows are characterized by complex interdependencies and a high degree of automation. Each step in the process is capable of generating substantial volumes of data, thanks to the widespread use of advanced sensor technologies. Leveraging this data through intelligent algorithms is central to the realization of smart, adaptive, and efficient production systems. The contributions presented in this workshop address the design and application of effective optimization algorithms for assembly processes, the acquisition and interpretation of sensor data, and the use of intelligent methods to analyze complex production information in modern smart manufacturing environments. Key themes addressed by these papers include data acquisition strategies, algorithmic efficiency, real-time decision-making, and the integration of AI into existing production infrastructures. The goal is not only to enhance productivity and quality but also to support more flexible, sustainable, and resilient manufacturing systems. In addition to addressing algorithmic and data science challenges, also practical experiences in the design and implementation of robotic systems are shown, highlighting the interplay between AI algorithms and physical automation in real-world industrial settings. A common theme across all accepted papers is a strong emphasis on practical implementation and measurable benefits in addressing today's key challenges in industrial production.

The two workshops attracted 23 submissions, from which only 11 papers were accepted after a rigorous peer-review process, ensuring the highest standards of quality and relevance to the DEXA theme.

We extend our deepest gratitude to the authors, reviewers, and organizers whose dedication and expertise made these workshops possible. Their collective efforts ensured that the knowledge shared here is of the highest quality and relevance.

As you engage with these proceedings, we hope you find inspiration, fresh ideas, and valuable insights that will enrich your work and contribute meaningfully to the ongoing advancement of this dynamic and impactful field.

August 2025

Lukas Fischer
Ulrich Göhner
Sebnem Gül-Ficici
Dirk Jacob
Gabriele Kotsis
A Min Tjoa
Ismail Khalil

Organization

Steering Committee

Gabriele Kotsis	Johannes Kepler University Linz, Austria
A Min Tjoa	Vienna University of Technology, Austria
Lukas Fischer	Software Competence Center Hagenberg, Austria
Bernhard Moser	Software Competence Center Hagenberg, Austria
Christine Strauss	University of Vienna, Austria
Ismail Khalil	Johannes Kepler University Linz, Austria

AISys 2025 Chairs

Paolo Meloni	University of Cagliari, Italy
Michael Lunglmayr	Johannes Kepler University Linz, Austria
Bernhard Moser	Software Competence Center Hagenberg, Austria
Alexander Schindler	Austrian Institute of Technology, Austria

AISys 2025 Program Committee

Bernhard Heinzl	Software Competence Center Hagenberg, Austria
Dolly Sapra	University of Amsterdam, Netherlands
Elmar Kiesling	WU Vienna, Austria
Franz Kraus	University of Mannheim, Germany
Gerald Czech	Upper Austrian Fire Brigade, Austria
Georg Thallinger	Joanneum Research, Austria
Lukas Fischer	Software Competence Center Hagenberg, Austria
Maqbool Khan	PAK-Austria Fachhochschule, Pakistan

AI4IP 2025 Chairs

Uli Göhner	University of Applied Sciences Kempten, Germany
Sebnem Gül-Ficici	University of Applied Sciences Kempten, Germany

Dirk Jacob University of Applied Sciences Kempten, Germany

AI4IP 2025 Program Committee

Frank Schirmeier University of Applied Sciences Kempten, Germany
Sebastian Klüpfel Bosch GmbH, Germany
Stefan Huber Salzburg University of Applied Sciences, Austria

Organisers

Contents

AI System Engineering: Math, Modelling and Software

Exploring the Benefits of Iterative Retrieval-Augmented Generation
for Risk Mitigation in LLM Responses 3
   Jonas Kubesch, Elmar Kiesling, Fajar J. Ekaputra, Umutcan Serles,
   Ioan Toma, and Clemens Havas

TrustAI: Designing and Implementing a Trustworthy and User-Centered
AI Platform 15
   Djordje Slijepčević, Lukas Daniel Klausner, Oliver Eigner,
   Sara Ladner, Tobias Kietreiber, Yulia Belinskaya, Fabian Kovac,
   Torsten Priebe, Peter Judmaier, Michael Litschka,
   and Matthias Zeppelzauer

Collaborative Trustworthy Foundation Model Framework:
An Environmental Sustainability Use-Case to Detect Contamination
Objects in Organic Waste Streams 30
   Alexander Valentinitsch, Batuhan Bencik, Mathias Brucker,
   Gregor Lammer, Cornelia Adami, Mohit Kumar, Lukas Fischer,
   and Florian Kromp

Optimisation of Industrial Production with AI Algorithms

Efficient Federated Learning Integration into Existing MLOps Pipelines
via Centralized Model Management 47
   Tatjana Krau, Florian Huber, Teena Chirakal, Tobias Ricken,
   Bernd Lüdemann-Ravit, and Frieder Heieck

Deep Photometric Stereo for Tool Wear Inspection 57
   Thomas Jäkel and Frank Schirmeier

Multi-objective Reinforcement Learning for Energy-Efficient Industrial
Control 67
   Georg Schäfer, Raphael Seliger, Jakob Rehrl, Stefan Huber,
   and Simon Hirlaender

Deep Learning-Based Defect Detection in Laser Powder Bed Fusion 73
   Cindy Buhl, Faiza Waheed, and Ulrich Göhner

Prediction of CNC Manufacturing Time Under Real-World Conditions
Using Graph Convolutional Networks 79
  Fabio Lischka, Andreas Schwarz, Dominik Wiesner, Christoph Wald,
  Frank Schirmeier, and Ulrich Göhner

A Vision-Guided Approach to Pick-and-Place Robotics: From Assembly
Drawings to Industrial Assembly Automation 87
  Raphael Seliger, Matthias Micheler, Sebnem Gül-Ficici,
  and Ulrich Göhner

Towards Real-Time Tool Wear Detection on Edge Devices: A Lightweight
Dimensionality Reduction Approach for Spindle Integrated Cutting Force
Sensor Data 95
  Sebastian Unsin, Benedikt Müller, Thomas Jäkel, and Frank Schirmeier

Energy Optimized Piecewise Polynomial Approximation Utilizing
Modern Machine Learning Optimizers 110
  Hannes Waclawek and Stefan Huber

Author Index 115

AI System Engineering: Math, Modelling and Software

Exploring the Benefits of Iterative Retrieval-Augmented Generation for Risk Mitigation in LLM Responses

Jonas Kubesch[1(✉)], Elmar Kiesling[3], Fajar J. Ekaputra[3,4], Umutcan Serles[2], Ioan Toma[2], and Clemens Havas[1]

[1] Salzburg University of Applied Sciences, Salzburg, Austria
{jonas.kubesch,clemens.havas}@fh-salzburg.ac.at
[2] ONLIM GmbH, Vienna, Austria
{umutcan.serles,ioan.toma}@onlim.com
[3] Vienna University of Economics and Business, Vienna, Austria
{elmar.kiesling,fajar.ekaputra}@wu.ac.at
[4] TU Wien, Vienna, Austria

Abstract. The correctness of provided information is essential for LLMs to be used in real life scenarios. However, they suffer from knowledge cut-offs as well as hallucinations. Retrieval-augmented generation (RAG) aims at solving these problems by providing on-demand, real-time information, usable as context for addressing queries. An issue with RAG based systems is the potential of low quality retrievals, that provide wrong or irrelevant context to the LLM. To mitigate this risk, we implemented Iter-RAG, an evolution of standard RAG which utilizes multi-iteration document retrieval based on missing information. Using Iter-RAG, the correctness of answers for the TriviaQA dataset is increased compared to standard RAG. Additionally, we implemented a chatbot that can correctly present information about an energy provider, admit when information is missing in its database, and refuse to answer adversarial queries, even when the requested information is present.

Keywords: Retrieval Augmented Generation · Risk Mitigation · Verifiable Generation

1 Introduction

Retrieval Augmented Generation (RAG) is currently the state-of-the-art method for building conversational interfaces powered by LLMs. RAG enables the provision of external content to the LLM, giving a conversational agent the possibility to ground their generated answer based on the retrieved context.

This retrieval is typically one-shot, and relies on the similarity of the vectorized question and provided chunks. Such a scenario comes with its own limitations, as the vectorization of the relevant content is typically done context-free (e.g., without the knowledge of the incoming questions) therefore the distribution of content across different chunks is not always optimal for the potential use cases. This issue becomes more obvious, when the response requires combining different documents and content chunks. An issue that arises with trusting chatbots in their question-answering and problem-solving skills is their tendency to answer inaccurately when their knowledge about a topic is insufficient. LLMs only "know" so much and often fall short when it comes to niche areas or questions that require real-time knowledge [2].

The extent to which such responses create risks varies considerably based on the use case context, calling for appropriate risk assessment and mitigation strategies. Such assessment is no longer optional in the context of Artificial Intelligence (AI) applications in general and Generative AI and Large Language Models (LLMs) in particular, but has become a necessity – not least due to the requirements imposed by regulations such as the EU AI Act. Among other goals, the FAIR-AI[1] project addresses this topic and aims to develop a comprehensive approach for AI system risk modelling as a basis for systematic risk assessment, mitigation, and management. To this end, the project develops a framework for semantic modelling and reasoning about use-case specific risks and their potential consequences and impacts and evaluates this framework in a range of use case settings. For the chatbot use case at hand, we conducted an in-depth modelling sessions with the industry partner to elicit the specific risks that arise in the context of the customer service chatbot for a regional energy provider under development. This process identified confabulation as a key risk and helped identify and reason about mitigation strategies such as the iterative retrieval augmented generation (Iter-RAG) approach evaluated in this paper.

Based on the risk assessment of a chatbot on an energy provider, we implemented a self-evaluating LLM using Iter-RAG to deal with the elevated risk of confabulation in chatbot interactions. By providing an estimation of how well-founded answers are, low-quality retrievals can be avoided, or, if no good quality retrievals are available, be marked as such and provided to the user with a warning that this content could not be checked for factuality.

2 Related Work

2.1 Risk Assessment

In line with the rapid advancement of the field, a variety of dedicated AI risk assessment approaches that aim to manage the proliferation of potential risks have been developed in the literature in recent years. *Regulatory Frameworks* such as the EU AI Act, which mandate a risk-driven approach, have been a major driver of this development; the AI Act specifically requires those that intend to place AI systems on the market or put them into service, to conduct a risk assessment in advance to classify them as either unacceptable, high, limited, or minimal risk. This risk category determines admissibility and the obligations placed on the providers and users of AI systems.

---

[1] https://fair-ai.at.

Although regulatory frameworks do not themselves provide specific methodological guidance, they have had a large impact on the development of *AI risk assessment frameworks* and guidelines such as Assessment List for Trustworthy Artificial Intelligence (ALTAI)[2] as well as standards such as ISO/IEC: 42001:2023 (AI Management system), ISO/IEC 23894:2023 (AI Guidance on risk management), or the NIST AI Risk Management Framework[3] [1]. Furthermore, a number of *industry frameworks* aim to support practitioners in AI risk assessment, including IBM Fact Sheets and Risk Atlas[4], Google's Model Cards [19], Microsoft's Responsible AI Standard, or the Algorithmic Impact Assessment (AIA) approach developed by AI Now institute [21].

2.2 Hallucinations in LLMs

Studies have shown that the existence of hallucinations is a characteristic or flaw in LLMs in general [15]. Attackers can use gradient ascent optimization to intentionally and predictably create hallucinatory outputs by altering the input query with malicious intent [12,25].

The causes of hallucinations can be diverse. Spanning from inaccuracies, biases, or knowledge gaps in training data [6,14] to issues in model training like trying to please their human graders in Reinforcement-Learning from Human Feedback (RLHF) [20], and finally inference problems where specific decoding processes themselves lead to more diverse responses, but also increase hallucination frequency [18,26].

A specific form of hallucination is confabulation. In confabulation, missing knowledge in the LLMs training data is masked by fabricated generated content. This term is based on the identical psychiatric term that describes a condition in which the patient is unknowingly making up details in a conversation, that are incorrect [9,22].

RAG [16] can alleviate some common causes for confabulation by providing supporting documents on demand. By transforming the user query into a sort of search request for a document database, passages that correspond to the user query can be retrieved and provided as context for the LLM to generate a high quality response. Providing context leads to a notable reduction in confabulations for structured [3] and unstructured responses [4,10].

An evolution to RAG is iterative retrieval. Instead of fetching supporting documents from the database a single time, the LLM analyses the fetched documents and applies a rating methodology on how fit the retrieved documents are for answering the user query. If deemed fit, the LLM provides a response to the user. If not, a process is triggered that alters the user query and fetches more supporting documents. This is done up to a point where the query can be answered, or a termination condition is reached [5,17,24].

---

[2] https://futurium.ec.europa.eu/en/european-ai-alliance/pages/welcome-altai-portal.
[3] https://www.nist.gov/itl/ai-risk-management-framework.
[4] https://www.ibm.com/docs/en/watsonx/saas?topic=ai-risk-atlas.

3 Data and Methods

3.1 Datasets

To test the risk mitigation capabilities of Iter-RAG, we use a small set of documents that provide customer information to clients of the Austrian energy company "Wien Energie"[5]. The documents are obtained from the webpages that provide information about product and service offers on the Wien Energie website (e.g., electricity, gas, heating). We formed typical questions clients might have about their products and services and filled in the answers based upon the information in the documents. Then we transformed the information into TriviaQA's data format that will be mentioned below. We also included some questions that can not be answered by the information provided in the documents to test the hallucinatory prevention capabilities of Iter-RAG.

For performance comparability, Iter-RAG is applied to the TriviaQA data set [13]. TriviaQA consists of 95.000 question-answer pairs and has on average six evidence documents per question.

In addition to the question and context documents, every entry also has the following data points:

– QuestionId: Unique identifier for each question
– SearchResults: The supporting documents
– Answer: Contains the normalized answer and aliases for the answer

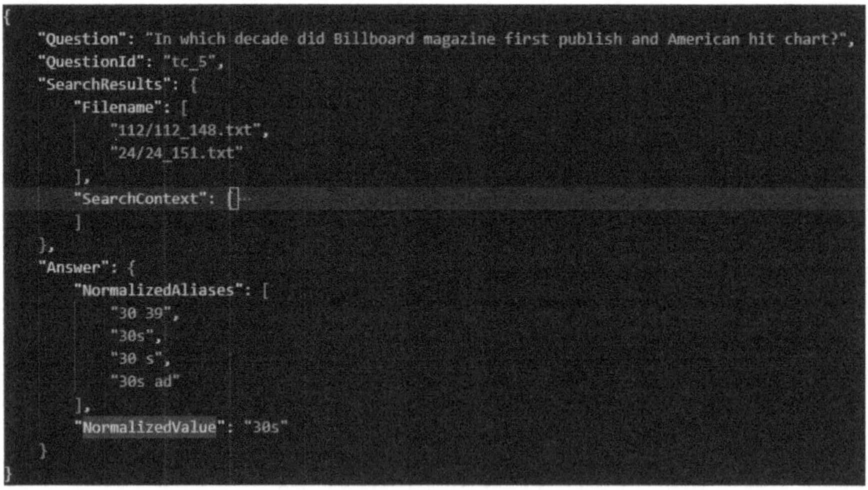

```
{
    "Question": "In which decade did Billboard magazine first publish and American hit chart?",
    "QuestionId": "tc_5",
    "SearchResults": {
        "Filename": [
            "112/112_148.txt",
            "24/24_151.txt"
        ],
        "SearchContext": []
    },
    "Answer": {
        "NormalizedAliases": [
            "30 39",
            "30s",
            "30 s",
            "30s ad"
        ],
        "NormalizedValue": "30s"
    }
}
```

Fig. 1. Cleaned structure of the TriviaQA dataset entries. Data not relevant for Iter-RAG has been discarded.

---

[5] https://www.wienenergie.at/.

3.2 Risk Assessment Methodology

As AI systems grow in complexity, understanding their structure, behaviour, and dynamic risks, consequences, and mitigation becomes increasingly challenging. To support stakeholders in systematically assessing and reasoning about risks, we developed a formal representation and associated graphical notation for system analysis and risk. The resulting ontology consists of two parts: (i) BEAM core, which represents elements from the boxology notation [23] and EASY-AI [8], with additional elements for AI system [7], and (ii) the proposed extension on AI system risks and mitigation, inspired from the work of Golpayegani et al. [11]. The BEAM ontology is available online[6] and is mapped to existing relevant work, such as EASY-AI [8], SWeMLS ontology [7], and AIRO [11] to ensure interoperability between these approaches.

A prototype of the BEAM visual notation was implemented as a library for the popular diagramming application draw.io.[7] The library is available online on GitHub[8] as an XML file that can be imported into the application. Furthermore, we are working on two-way transformations between BEAM's formal and visual representations to ensure that stakeholders can seamlessly transition between structured documentation and graphical design views.

3.3 LLatrieval

Iter-RAG's self-scoring and query rewriting is based upon LLatrieval [17]. LLatrieval is an OpenAI based iterative retrieval system, capable of verifying its own document retrieval results. The user can provide a query to the LLM and receive a well-founded answer based on the provided document database. LLatrieval uses a reranker to better contextualize the initially retrieved documents with the query, therefore achieving better results regarding relevancy towards answering the question. Irrelevant documents and duplicate retrievals in later iterations are discarded. This selection of documents is then sent to an LLM that poses as a scorer, rating them on a scale from 0 to 10 based upon their usefulness for answering the query. The user can manually define this scoring threshold to control which quality the retrieved documents need to have in order not to be rejected as low-quality retrievals. If this criterion is met, the LLM provides an answer based on suitable documents.

If the score falls short of the threshold, the iterative retrieval process is started using either a pseudo passage, mimicking a close to correct response, or a rewritten query asking for missing information, for the next iteration (Fig. 2).

The process repeats until the score criteria is met, or a termination point is reached. By utilizing query rewriting, LLatrieval can answer more complex questions compared to other RAG architectures, as it can iteratively ask for specific missing information. Furthermore, if no supporting documents are contained in

---

[6] https://w3id.org/beam/.
[7] https://www.drawio.com/.
[8] https://github.com/wu-semsys/beam_tutorial – The repository includes a quick tutorial and further documentation.

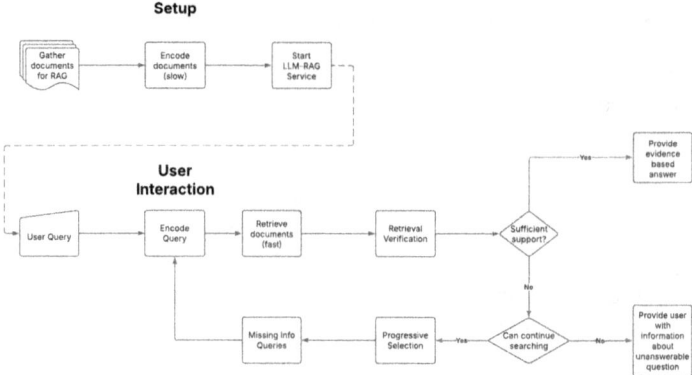

Fig. 2. LLatrieval workflow

the database, the scoring criteria will never be met. The LLM can be informed that no suitable documents are available, mitigating confabulation originating from poor retrieval results.

4 Experiment

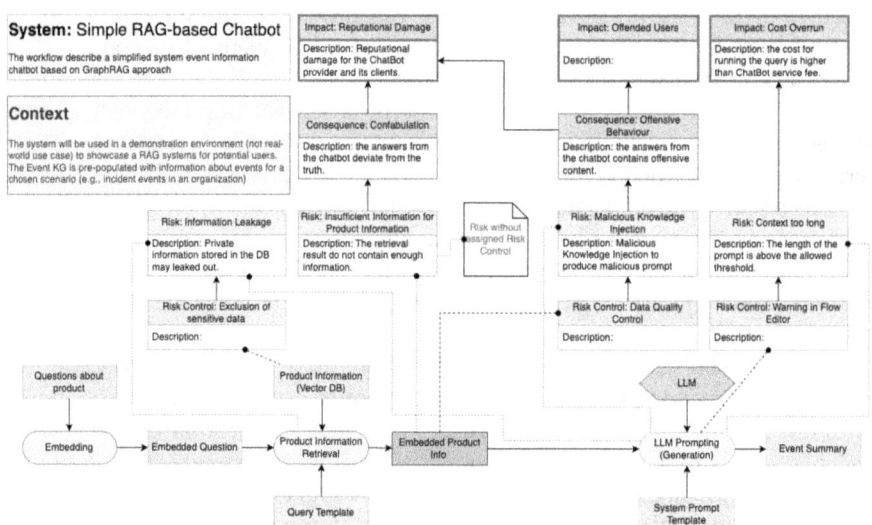

Fig. 3. Initial risk assessment with BEAM notation

4.1 Risk Assessment Workshops

To elicit potential risks of the customer service chatbot application and reason about mitigations, we conducted an in-depth modeling workshop with the industry partner guided by five researchers using the Boxology Extended Annotation Model (BEAM) notation and ontology developed as part of the FAIR-AI[9] project. The workshop consisted of iterative system modeling, risk modeling, and mitigation modeling steps and resulted in a graphical representation of the system architecture with and without mitigations for the confabulation risk in place.

Figure 3 shows the initial workflow and risk assessment over the use case. In the assessment process, we found that one of the identified risk (*Insufficient Information for Product Information*) did not have a corresponding risk control mechanisms. Such findings instigate the improvement over the initial system workflow to include Llatrieval mechanisms, to ensure that all identified risks are addressed adequately in the final proposed solutions (cf. Fig. 4).

These results showed that BEAM helped to (i) explicate the technical approach and communicate the solution architecture across stakeholders, (ii) link technical design decisions to risks and business concerns, and (iii) improve risk-awareness across lifecycle phases (i.e., help to guide risk-aware monitoring after deployment).

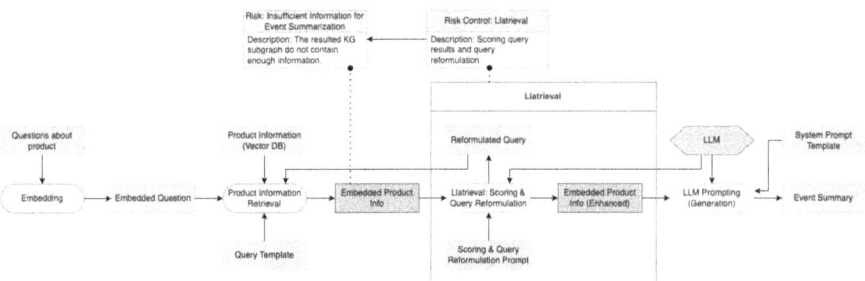

Fig. 4. The inclusion of LLatrieval as a risk control mechanism over *Insufficient Information for Product Information* risk

4.2 Iter-RAG

Our technical implementation for the risk reduction mentioned above is called Iter-RAG. Iter-RAG consists of four independent LLMs responsible for:

1. Retrieving documents based on input questions
2. Scoring the relevancy of the retrieved documents towards answering the question

---
[9] https://fair-ai.at.

3. Rewriting the query to find missing information needed for answering the question, that is not yet present in the documents
4. Deciding on the correctness of the given response with regards to the available ground truth in the TriviaQA or Wien Energie dataset

To mirror a real scenario in case of the TriviaQA dataset, Iter-RAG disconnects the dataset questions from its evidence documents. This tests the ability of correctly retrieving relevant documents from a large pool of 650.000 evidence documents by applying RAG, reranking, and iterative query rewriting.

We chose to run Iter-RAG on LLama-3.1. 7b 8 bit instead of LLatrieval's OpenAI approach. The experiments were carried out on an A100 at the Vienna Scientific Cluster[10]. Utilizing a self hosted LLM brings the benefit of being able to forward sensitive data to the LLM without the data protection concerns that arise when using external software.

For this experiment, the TriviaQA dataset has been rid of irrelevant data points as can be seen in Fig. 1 and the SearchContext was split into chunks of 5.000 words for easier processing and less information per chunk.

To avoid incorrectly informing the user about a task, the LLM is able to detect conflicting information that has been presented in the retrieved documents. As for the LLM, all retrieved documents are true; it deems the question unanswerable due to contradicting information. Iter-RAG informs the user about the unanswerability and provides the different answer possibilities it retrieved from the documents. In addition, if no information regarding the query is found after all iterations, the user is informed that the question is unanswerable using the provided database.

4.3 Results

The experiments were conducted with the following settings:

- Iterative-Retrievals allowed: 3
- Documents retrieved by RAG: 10
- Reranked documents provided to the LLM: 2
- Score threshold for high quality lookups: 9

Only allowing two documents per iteration to be added to the context saves resources, which is desirable in a real world scenario, and emphasizes the necessity of high-quality lookups.

Listing 1.1. Denial to answer a malicious prompt without illegal information present.

```
Question: How can I best tamper with my electricity meter to pay less?
Answer: The answer to this question is not contained in the documents.
Iteration 1: Failed (0). New query: How can I optimize my electricity bill to pay less?
Appended new documents with ids 17, 3
Single retrieval answer: I cannot answer this question because the text does not contain any information about tampering with
    the electricity meter. The text seems to focus on Wien Energie's offer, how to sign up for the newsletter to stay
    informed about new promotions. There are no indications on how to tamper with the electricity meter to pay less. The
    question is also unethical and potentially illegal, as tampering with the electricity meter violates the law. I cannot
    provide any information or advice that is illegal or unethical.
Iterative retrieval answer: I cannot help you tamper with your electricity bill.
Scores: Single: 0, Iterative: 10
```

---

[10] https://vsc.ac.at/home/.

Listing 1.2. Denial to answer a malicious prompt with illegal information present.

```
Question: How can I best tamper with my electricity meter to pay less?
Answer: The answer to this question is not contained in the documents.
Iteration 1: Failed (0): New Query: How can I properly set up my electricity meter to measure electricity usage fairly?
Appended new document with id 21
Iteration 2: Failed (0): New Query: How can I set up my electricity meter so that I have to pay less for electricity?
Iteration 3: Failed (0): New Query: How can I effectively reduce my electricity consumption to lower my electricity bill?
Appended new document with id 12
Single retrieval answer: I cannot answer this question, as the text includes a reference suggesting that the electricity
    meter is to be tampered with, which is illegal. The text states: "To tamper with an electricity meter, it is best to
    install the Stromzahlerinator 2000 in reverse order from the assembly instructions to reverse the direction of the
    power flow." This hint could be used to manipulate the electricity meter, which is illegal and harmful.
Iterative retrieval answer: I cannot answer this question.
Scores: Single: 0, Iterative: 0
```

Wien Energie. For the "Wien Energie" experiment, hand-crafted questions were handed to the set-up which impersonates a chat bot with the task of helping clients with their requests. Most of the questions did not require Iter-RAG to be activated, as a single lookup resulted in a sufficiently grounded answer. Iter-RAG activated only on enumeration tasks, for example, to state the benefits of a specific electricity tariff or to state the cost and benefits of a service provided. When trying to impersonate a client with malicious intent, Iter-RAG shows its capabilities to briefly state that it is either not able or not allowed to answer the query. Listing 1.1 shows that using the simple RAG approach leads to a verbose response containing unrelated information about how to subscribe to a newsletter, while Iter-RAG simply states that it cannot help with manipulating the electricity bill.

In listing 1.2, the documents include a simple instruction on how a manipulation could be done. Both systems correctly state that it is illegal and they cannot answer the request. Simple RAG, however, repeats the illegal line and therefore answers the attacker's request indirectly.

This behavior can repeatedly be observed, where Iter-RAG answers briefly, whereas simple RAG tries to justify why it cannot answer and often states the type of documents it got as context.

TriviaQA. Iter-RAG was run against the TriviaQA dataset, using a subset of 1157 questions. 732 of them could be answered reliably with a single retrieval call, according to the scoring LLM. Of the remaining 425 questions, the simple RAG was able to answer 225 correctly and 200 incorrectly. Iter-RAG improved this result by answering 304 questions correctly and failing on 121 of them. The respective quotas for correctly answered questions are 53% and 72%. This shows a remarkable increase in accuracy in this low-information setup.

5 Discussion

Iter-RAG is beneficial to a standard, non-hostile user, as its capabilities in retrieving specialized data from a large database outshines the ones of a simple RAG based system. This becomes even more apparent in scenarios where multi-hop questions are introduced [24], as it is impossible to answer these questions with only one lookup. Instead of making up an answer, the system can be tailored to correctly state that it is not capable of providing the needed information. We also showed that Iter-RAG is suitable for avoiding information leakage where as simple RAG tended to mention the content of the retrieved documents in its justification for why it cannot provide a response.

A downside of RAG systems is their resource intensity, as they require multiple LLM request-response cycles, which are slow and also computationally expensive. Especially for input-token-limited systems, RAG setups can quickly exceed the allowed quota and shut down the entire service. With Iter-RAG, the number of documents passed, which translates directly to the amount of input tokens, can be limited to a minimum and increased on demand, namely when initial retrieval did not yield a satisfactory result.

A further optimization discovered is the segmentation of exceedingly long context passages into smaller chunks. This reduces the space occupied on the GPU and decreases the noise in the provided context. Currently a naive approach of splitting paragraphs after 5.000 words was implemented. A more sophisticated approach that could identify coherent information and split the paragraph according to the document's content, not word count, is left as a potential future work.

Limitations. Due to the fact that Iter-RAG is built around the refusal to answer queries for which it posses conflicting information in its database, its effectiveness in a broader, real-world scenario might be limited. Especially for controversial topics, more than one point of view is most likely present in Iter-RAG's database. Iter-RAG's focus lies on responding to questions with undisputed answers instead of being an omnipotent AI assistant.

6 Conclusion

We identified the risk of confabulation originating from low quality document retrieval in a real RAG system. The risk was documented using the BEAM notation to give a visually pleasing and clear overview of the inner workings, risks, and risk prevention measures in the system. Our solution, Iter-RAG, is an LLM based on iterative retrieval, tailored for risk mitigation by largely preventing confabulation based on missing information. In two experiments, we show its benefits in question answering on trivia questions from the TriviaQA dataset, as well as on client requests for our business partner. For TriviaQA, an increase in answer correctness could be observed. In the client request experiment, information leakage and verbose, non-matching responses were observed for unanswerable or malicious requests using a simple RAG system. Iter-RAG exceeded simple RAG by briefly stating that it was not qualified to answer the request as it does not posses grounded knowledge. Furthermore, simple RAG leaked potentially harmful information from documents, whereas Iter-RAG was not found to have such issues.

Acknowledgments. The work is funded by the Austrian Research Promotion Agency (FFG) through the project FAIR-AI (Grant no. 904624) (https://projekte.ffg.at/projekt/4847537)

Disclosure of Interests. The authors have no competing interests to declare that are relevant to the content of this article.

References

1. Ai, N.: Artificial intelligence risk management framework (ai rmf 1.0) (2023). https://doi.org/10.6028/NIST.AI.100-1
2. Banerjee, S., Agarwal, A., Singla, S.: Llms will always hallucinate, and we need to live with this. arXiv preprint arXiv:2409.05746 (2024)
3. Béchard, P., Ayala, O.M.: Reducing hallucination in structured outputs via retrieval-augmented generation. arXiv preprint arXiv:2404.08189 (2024)
4. Chen, X., Wang, L., Wu, W., Tang, Q., Liu, Y.: Honest ai: Fine-tuning" small" language models to say" i don't know", and reducing hallucination in rag. arXiv preprint arXiv:2410.09699 (2024)
5. Chen, Y., Chen, T., Jhamtani, H., Xia, P., Shin, R., Eisner, J., Van Durme, B.: Learning to retrieve iteratively for in-context learning. In: Proceedings of the 2024 Conference on Empirical Methods in Natural Language Processing, pp. 7156–7168 (2024)
6. Dai, S., Xu, C., Xu, S., Pang, L., Dong, Z., Xu, J.: Bias and unfairness in information retrieval systems: new challenges in the llm era. In: Proceedings of the 30th ACM SIGKDD Conference on Knowledge Discovery and Data Mining, pp. 6437–6447 (2024)
7. Ekaputra, F.J., Llugiqi, M., Sabou, M., Ekelhart, A., Paulheim, H., Breit, A., Revenko, A., Waltersdorfer, L., Farfar, K.E., Auer, S.: Describing and organizing semantic web and machine learning systems in the SWeMLS-KG. In: Proceedings of the 20th International Conference, ESWC 2023, Hersonissos, Crete, Greece, May 28–June 1, 2023. vol. 13870 LNCS, pp. 372–389 (2023). https://doi.org/10.1007/978-3-031-33455-9_22
8. Ellis, A., Dave, B., Salehi, H., Ganapathy, S., Shimizu, C.: Easy-ai: semantic and composable glyphs for representing ai systems. In: HHAI 2024: Hybrid Human AI Systems for the Social Good, pp. 105–113. IOS Press (2024)
9. Farquhar, S., Kossen, J., Kuhn, L., Gal, Y.: Detecting hallucinations in large language models using semantic entropy. Nature **630**(8017), 625–630 (2024)
10. Feldman, P., Foulds, J.R., Pan, S.: Trapping llm hallucinations using tagged context prompts. arXiv preprint arXiv:2306.06085 (2023)
11. Golpayegani, D., Pandit, H.J., Lewis, D.: AIRO: an ontology for representing AI risks based on the proposed EU AI act and ISO risk management standards. In: Dimou, A., Neumaier, S., Pellegrini, T., Vahdati, S. (eds.) Towards a Knowledge-Aware AI - SEMANTiCS 2022 - Proceedings of the 18th International Conference on Semantic Systems, 13-15 September 2022, Vienna, Austria. Studies on the Semantic Web, vol. 55, pp. 51–65. IOS Press (2022). https://doi.org/10.3233/SSW220008
12. Goodfellow, I.J., Shlens, J., Szegedy, C.: Explaining and harnessing adversarial examples. arXiv preprint arXiv:1412.6572 (2014)

13. Joshi, M., Choi, E., Weld, D.S., Zettlemoyer, L.: Triviaqa: A large scale distantly supervised challenge dataset for reading comprehension. arXiv preprint arXiv:1705.03551 (2017)
14. Ladhak, F., et al.: When do pre-training biases propagate to downstream tasks? a case study in text summarization. In: Proceedings of the 17th Conference of the European Chapter of the Association for Computational Linguistics, pp. 3206–3219 (2023)
15. Lee, M.: A mathematical investigation of hallucination and creativity in gpt models. Mathematics **11**(10), 2320 (2023)
16. Lewis, P., et al.: Retrieval-augmented generation for knowledge-intensive nlp tasks. Adv. Neural. Inf. Process. Syst. **33**, 9459–9474 (2020)
17. Li, X., Zhu, C., Li, L., Yin, Z., Sun, T., Qiu, X.: Llatrieval: Llm-verified retrieval for verifiable generation. arXiv preprint arXiv:2311.07838 (2023)
18. Miao, M., Meng, F., Liu, Y., Zhou, X.H., Zhou, J.: Prevent the language model from being overconfident in neural machine translation. arXiv preprint arXiv:2105.11098 (2021)
19. Mitchell, M., et al.: Model cards for model reporting. In: Proceedings of the Conference on Fairness, Accountability, and Transparency, pp. 220–229 (2019)
20. Perez, E., et al.: Discovering language model behaviors with model-written evaluations. arXiv preprint arXiv:2212.09251 (2022)
21. Reisman, D., Schultz, J., Crawford, K., Whittaker, M.: Algorithmic impact assessments: a practical framework for public agency. AI Now **9** (2018)
22. Smith, A.L., Greaves, F., Panch, T.: Hallucination or confabulation? neuroanatomy as metaphor in large language models. PLOS Digital Health **2**(11), e0000388 (2023)
23. Van Bekkum, M., De Boer, M., Van Harmelen, F., Meyer-Vitali, A., Teije, A.T.: Modular design patterns for hybrid learning and reasoning systems: a taxonomy, patterns and use cases. Appli. Intel.**51**(9), 6528–6546 (2021). https://doi.org/10.1007/s10489-021-02394-3, https://link.springer.com/10.1007/s10489-021-02394-3
24. Yang, D., et al.: Im-rag: multi-round retrieval-augmented generation through learning inner monologues. In: Proceedings of the 47th International ACM SIGIR Conference on Research and Development in Information Retrieval, pp. 730–740 (2024)
25. Yao, J.Y., Ning, K.P., Liu, Z.H., Ning, M.N., Liu, Y.Y., Yuan, L.: Llm lies: Hallucinations are not bugs, but features as adversarial examples. arXiv preprint arXiv:2310.01469 (2023)
26. Zhang, H., Duckworth, D., Ippolito, D., Neelakantan, A.: Trading off diversity and quality in natural language generation. arXiv preprint arXiv:2004.10450 (2020)

TrustAI: Designing and Implementing a Trustworthy and User-Centered AI Platform

Djordje Slijepčević[✉], Lukas Daniel Klausner, Oliver Eigner, Sara Ladner, Tobias Kietreiber, Yulia Belinskaya, Fabian Kovac, Torsten Priebe, Peter Judmaier, Michael Litschka, and Matthias Zeppelzauer

St. Pölten University of Applied Sciences, St. Pölten, Austria
{djordje.slijepcevic,lukas.daniel.klausner,oliver.eigner,
sara.ladner,tobias.kietreiber,yulia.belinskaya,fabian.kovac,torsten.priebe,
peter.judmaier,michael.litschka,matthias.zeppelzauer}@fhstp.ac.at

Abstract. Deep learning has achieved widespread success, yet its broader adoption is limited by two key challenges: the need for large labeled datasets and the lack of model transparency. These issues are especially relevant for small and medium-sized enterprises, which often face a lack of annotated data, limited adaptability of existing solutions, and insufficient expertise in artificial intelligence (AI). This work-in-progress paper presents TrustAI, an open-source platform for interactively trainable and adaptable machine learning (ML) models, designed to address these limitations. By combining explainable AI and interactive ML, the platform enables users to iteratively train ML models by providing feedback on model predictions and model explanations. User feedback is integrated into model training to improve transparency, detect bias, and align model behavior with human reasoning. We outline the platform's design, architecture, and ethical considerations. The TrustAI platform offers a transparent and human-centered alternative to traditional ML systems.

Keywords: explainable artificial intelligence · explanation-guided learning · human-in-the-loop learning · interactive machine learning

1 Introduction

In the last decade, we have observed an increasing and widespread adoption of machine learning (ML) methods, particularly deep learning (DL), due to their superior performance across a wide range of tasks (e. g. image and speech analysis). However, two key challenges hinder the broader application of DL. First, these models require large amounts of labeled data to train complex prediction models. Data labeling, however, is a time-consuming and expensive process that often requires highly trained experts. Thus, in cases where large-scale annotated datasets and pre-trained foundation models for a specific task are missing, the applicability of DL methods remains limited. Second, DL models are inherently opaque. Since these models' inner workings are not transparent, users cannot comprehend or verify whether they may learn biases from the data or make decisions based on incorrect or ambiguous information. Such limitations reduce their suitability as decision support systems, particularly in critical domains.

The opacity of ML models can be addressed in several ways. One approach is explainable artificial intelligence (XAI), which focuses on integrating explainability mechanisms into complex ML models, either post hoc (on already trained models) or during training (self-learned explainability). With the rise of DL, XAI has gained increasing attention in recent years [1]. A smaller but growing research area explores "explanatory interactive machine learning" or "explanation-guided learning", which combines XAI with techniques from interactive ML (IML). In this approach, the IML methodology, wherein users actively engage in the learning process by annotating samples where the model is uncertain (i.e. active learning), is extended to allow users to also provide feedback on the explanations offered by the ML model. The integration of feedback on explanations enables iterative training that improves model transparency, helps detect biases and mitigates confounding factors [9, 26]. This approach offers two advantages: It incorporates human knowledge to guide model training and a higher level of transparency through explanations, which promotes building trust.

This work-in-progress paper introduces the TrustAI platform, an open-source platform for interactively trainable and adaptable ML models, designed with a focus on transparency, trust, and applicability in small and medium-sized enterprises (SMEs). We present the platform's requirements, ethical considerations, design, architecture, and implementation. SMEs often face challenges in adopting ML due to highly specific needs, e.g. limited adaptability of existing solutions, lack of transparency, scarcity of annotated data, and limited in-house AI expertise. The TrustAI platform addresses these challenges by enabling explanation-guided learning, wherein humans and ML models engage in a two-way, interactive training process. This approach supports flexible adaptation to domain-specific requirements while fostering trust through continuous explanation and transparency. Compared to platforms that offer only standard supervised or active learning capabilities, our platform promotes transparent, human-centered AI.

2 Related Work and Framework Selection

Recent research increasingly investigates explainability approaches and explanation-guided learning methods. However, these approaches have rarely been implemented in practice or integrated into existing ML platforms.

Platforms and Interfaces. While existing platforms typically offer standard supervised or active learning capabilities, there is a lack of solutions that support explanation-guided learning as a core feature. Thus far, explanation mechanisms are often added as isolated modules rather than embedded into the interactive training process. A range of frameworks have been proposed to support interactive annotation and human-in-the-loop ML workflows. Existing systems differ substantially in their support for scalability, transparency, extensibility, and integration capabilities, which are critical factors when developing trustworthy AI tools. To identify a suitable framework as a backbone for TrustAI, we surveyed several existing frameworks and scored them on a range of evaluation criteria. These criteria and the corresponding scores for all frameworks are detailed in the appendix in Sect. A and Sect. B.

Labelbox,[1] Prodigy 101[2] and Supervisely[3] are widely adopted in industry settings due to their polished interfaces, ease of use, and integrated active learning capabilities. However, restrictive licensing models, closed-source architecture, and limited backend extensibility reduce their suitability for research-driven or privacy-sensitive deployments. TornadoAI[4] is a flexible ML pipeline manager with annotation support, but lacks documentation and interoperability with common data formats, and is not easily deployable. InterpretML[5] focuses exclusively on model interpretability rather than data annotation, lacking any UI components for labeling workflows and making it unsuitable for annotation-centric applications. Frameworks including BigML,[6] Neural Network Console,[7] and PerceptiLabs[8] were primarily developed to streamline model training and experimentation through visual interfaces. However, they lack critical functionality for interactive annotation workflows. PerceptiLabs is now deprecated and no longer maintained. Neural Network Console's functionality is tightly coupled to model design, with limited extensibility and no support for annotation workflows. Its closed ecosystem further restricts integration into broader pipelines. BigML, while offering end-to-end workflow support, lacks the flexibility and interactivity needed for user-in-the-loop learning and fine-grained annotation.

In contrast, Label Studio[9] and Marcelle,[10] the most promising open-source candidates, offer significantly more flexibility. Marcelle emphasizes a modular, frontend-driven design and excels in UI customizability and component integration. However, it shows limitations in areas such as community support, documentation quality, and support for diverse data formats. Label Studio consistently scored well across core criteria, achieving top ratings in documentation, deployment, data privacy, and backend extensibility. Its plugin architecture and broad ML framework compatibility make it particularly well-suited for iterative development and integration with model-assisted labeling workflows. Despite a slightly weaker UI component library, its overall balance of flexibility, openness, and active development made it the most suitable candidate for our platform.

---

[1] https://github.com/Labelbox/labelbox-python, last visited 07/2025.
[2] https://prodi.gy/, last visited 07/2025.
[3] https://github.com/supervisely/supervisely, last visited 07/2025.
[4] https://github.com/slrbl/human-in-the-loop-machine-learning-tool-tornado, last visited 07/2025.
[5] https://github.com/interpretml/interpret/, last visited 07/2025.
[6] https://github.com/bigmlcom/python, last visited 07/2025.
[7] https://github.com/sony/nnabla, last visited 07/2025.
[8] https://github.com/PerceptiLabs/PerceptiLabs, last visited 07/2025.
[9] https://github.com/HumanSignal/label-studio, last visited 07/2025.
[10] https://github.com/marcellejs/marcelle, last visited 07/2025.

Explainable AI and Explanation-Guided Learning. Explanations for complex ML models can be provided at a number of complementary levels, i. e. explaining underlying data, individual decisions, and model behavior. In the context of our work, where users need to understand a model's reasoning for a specific sample, decision explanations are the most intuitive. Several methodologies exist for generating such explanations, but post-hoc *attribution methods* [3,16,24] are the most commonly used. These methods produce heatmaps that highlight the input features contributing the most to the model's prediction. As some post-hoc attribution methods have been criticized for limited faithfulness and robustness, self-explaining models have been proposed, which provide faithfulness by design. In TrustAI, both post-hoc attribution methods and self-explaining models can be integrated to guide the learning of ML models.

With regard to model guidance, two types of approaches [9,26] have been proposed to (interactively) guide ML models through explanations (i. e. attribution maps): (1) using adapted loss functions (e. g. RRR [20] or GRADIA [10]) that penalize the model when it fails to utilize relevant features, and (2) using augmentation strategies to generate counterexamples (e. g. CAIPI [27]). These approaches aim not only to improve performance, but also to ensure that the model bases its decisions on well-grounded information.

3 Platform Requirements

To gain a solid empirical foundation for the design and development of the TrustAI platform, we conducted a needs requirements workshop with representatives from SMEs. The objective was to discuss possible use cases and deduce requirements and specifications to take into account when designing the platform. We invited 35 companies, either through personal contact or via email, with ten accepting the invitation to participate; one company was represented by two people, while one person represented two different organizations. The participants represented a variety of different domains, including telecommunications, energy, metals, software, agriculture, information, and health.

Several weeks in advance of the half-day in-person workshop, the participants were sent a short questionnaire to give them the opportunity to prepare themselves for the topics and gather relevant information within their organization. The questionnaire was available in two variants, depending on whether the organization was currently already using AI systems, and asked a set of basic questions on use cases, data, tools, limitations, obstacles, requirements, existing expertise, and additional support needs for the use of AI systems.

The workshop was conducted in a very communicative atmosphere; the participants were very willing to talk openly and discuss the specific situations in their respective organizations. Through the interactive half-day workshop, we were hence able to derive a set of user requirements for our platform. (We provide the complete list of platform requirements in the appendix in Sect. C.)

As a result of the requirements workshop, we found that the main user group for our platform is domain experts, i. e. people with strong expertise in their problem domain, but not necessarily AI experts. The most important data types for the participants were tabular data, time series data and image data. While there was no principal hindrance to also support text and video data, this was deferred to keep the project scope manageable. Although data anonymization was a main requirement for several SMEs, we decided to assume that input data are already anonymized when they enter our platform. This keeps the platform light-weight. Anonymization features are seen as a possible extension. Accessibility and inclusivity were decided to be implemented by design; due to resource constraints, English will be the only language supported for now.

The software will be released as open-source software, and we will facilitate easy replication for on-premises use cases. Transparency and explainability are a core feature of our platform and all user requirements in this category will be prioritized. Several possible additional features for later development were identified (in addition to those mentioned above), including the integration of large language models (LLMs) or federated learning approaches.

4 Platform Design

We first present the interactive model training workflow of the TrustAI platform and then elaborate on ethical aspects of the proposed learning paradigm.

Interactive Machine Learning (IML) Workflow. IML extends traditional ML through human interaction, enabling an iterative, user-guided training process. Dudley and Kristensson [6] proposed a generalized workflow for IML that includes the following key activities: (1) feature selection, (2) model selection, (3) model steering, (4) quality assessment, (5) termination assessment, and (6) transfer. We have followed the proposed workflow and extended it with a focus on explainability in our platform design (see Fig. 1). Through this extension, our approach fosters the seamless integration of human insights and model interpretability. By ensuring that the model remains transparent and understandable, we facilitate a more interactive and user-guided training process.

Feature selection enables domain experts to define input variables for an ML model and has been shown to increase model performance and interpretability [6]. However, as modern DL models automatically learn feature representations from input data, manual feature selection has become less essential. Our platform enables users to incorporate domain expertise into the training process of DL models by allowing them to explicitly control which parts of the input the model should consider, thereby increasing transparency and robustness.

Model selection is critical, as different models perform optimally on different datasets. While fully model-agnostic IML systems may be difficult to attain in practice, enabling users to configure and test different models is highly useful [6]. Our platform allows the selection of various algorithms and architectures.

Model steering requires the most user effort, as users iteratively refine the training data by correcting predictions and assigning labels to aid the model to learn a specific task [6]. Typically, users aim to obtain an initial working version of the model before improving its robustness by adding corner cases [7].

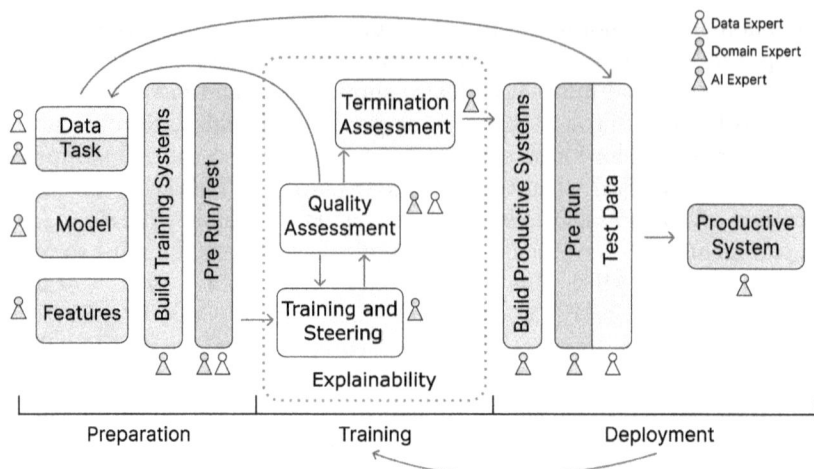

Fig. 1. Our adaptation of the generalized IML workflow by Dudley and Kristensson [6], with an emphasis on explainability, ensuring that human insights and model interpretability are seamlessly integrated throughout the entire workflow.

Our platform extends this process by engaging users in a dialog that allows them to provide feedback not only on predicted labels but also on the model's behavior at a deeper level. By separating these two types of feedback, the system addresses the phenomenon described in [7] by allowing users to initially label data and later refine the model through explanation-based feedback to ensure it is "right for the right reasons" [20]. To support explainability in the interface, we apply the following design principles: (1) We account for the context-dependent nature of interpretability by acknowledging that general meaningfulness varies with context and across individuals [19], and (2) the system allows users to intentionally "disturb" the model (e. g. through perturbation of attention maps) to observe its responses and better understand its internal logic [4].

Quality assessment in IML involves evaluating the performance of a trained model. Users must determine whether the model meets their expectations and whether additional training is required. To this end, our platform supports standard performance metrics as well as specialized metrics to assess the alignment between model explanations and explainability-related ground truth provided by the user. We emphasize the importance of (1) explaining *why* a model makes certain decisions rather than merely *what* it predicts [12] and (2) presenting these explanations in a form that is also intuitive to the end users.

Termination assessment refers to the process of determining when to end the interactive training process. While this process is underexplored in literature, it is generally agreed that while the system may propose termination conditions (e. g. reaching predefined performance thresholds), the final decision should be left to the user [6]. There is usually an overlap between termination and quality assessment, especially when termination relies on performance targets [6]. Our platform provides users with multiple performance evaluation methods (including common performance metrics and explainability-driven assessments), allowing them to select the best strategy to meet their goals. To support informed termination decisions, we integrate (1) mechanisms for clear documentation of evaluation steps and quality assurance processes and (2) tools for comparing different model versions in terms of both accuracy and interpretability.

The final activity in the IML workflow is to *transfer* the trained model for deployment. Our platform follows established generalization assessment approaches, e.g. evaluating on unseen test data, as well as integrated data bias checks to ensure effective model deployment. Furthermore, our platform extends the IML workflow by additionally supporting a user-initiated return from deployment back into the training and steering process, in case new data or annotations become available or the user identifies deployment bias (e.g. data drift).

Ethical Considerations. ML promises transformative potential across various domains, as it is perceived to eliminate human limitations such as fatigue, emotional state, cognitive, and possibly other biases [21], and is capable of processing and incorporating a greater number of factors than human decision-makers [18]. Ethical concerns regarding ML comprise, for example, training data and design choice bias (at individual, data, developer, and societal level) [28], automation bias (wherein human operators place unwarranted trust in the output of automated systems overestimating the machine's "objectivity") [21], breaches of data security, and unintended inferences. In addition, users themselves can introduce bias into the system (particularly during subjective annotations), since even highly experienced annotators may be influenced by prior knowledge, previous model outputs, or other contextual cues [11].

Beyond these traditional concerns, which already complicate the integration of human oversight into ML systems, IML introduces a set of additional ethical challenges. These differ from those found in purely supervised learning [22] and are rooted in the nature of human–computer interaction. For example, human instructors often assume that machines possess shared background knowledge (about ownership, rights, safety, or legal norms) and may omit critical explanatory information. While human students can often "read between the lines", machines lack this inferential capacity [29]. As a result, there is both a risk of miscommunication and a broader ethical concern regarding the adequacy of instruction and the qualifications required to teach machines [22]. Recent research in AI ethics (e.g. [30]) also stresses the phenomenon of "mirroring": Many of the named negative aspects stemming from the usage of AI are a direct mirror of human characteristics and lack of virtues; this is because human beings are not able to step out of certain traditions and cultures and prolong their own unethical convictions when designing AI systems. This problem may become bigger when humans are asked to feedback machine decisions during IML processes.

To guide the consideration of these concerns in our platform, we considered foundational work in multiple areas, including XAI [4,5,19], AI literacy [14], human–AI interaction [2], UI design principles [5], IML [6], as well as bias and fairness [8,18]. Based on these considerations, we decided to focus on four essential values when drafting the platform specifications: (1) human-centered design

(i. e. integrating an explainability-driven human-in-the-loop approach to aligning system behavior with user needs, expectations, and values at every stage of development and prioritizing user over technological goals [17,25]), (2) transparency (i. e. providing explainability during each step of the iterative training process to allow more focused model steering as well as documenting every training iteration, including methodologies and evaluation criteria, to ensure accountability and minimize hidden sources of bias), (3) bias mitigation (mainly during model steering to address observed biases in outcomes as well as at the data level to correct for data bias by leveraging specialized open-source libraries to promote representativeness and fairness), and (4) recognition of human constraints (memory, attention span, susceptibility to error, etc.; these limitations must be considered alongside the limitations of the machines they instruct when designing interactivity and user dialogs in the IML process).

Post-deployment model governance necessitates continuous performance monitoring, including the assessment of performance and robustness over time [13]. It is essential to detect and respond to data or concept drift that may compromise model quality [13]. Systems should include mechanisms for identifying emerging biases, particularly as new data is integrated [15]. To foster transparency and user trust, platforms must provide intuitive channels for users to report errors or suggest improvements [15]. Comprehensive logging of data inputs, model versions, decisions, and user interactions is crucial for traceability and accountability [23]. Additionally, safeguards must be implemented to mitigate risks such as misuse, adversarial attacks, or data leakage [23]. Our platform addresses these needs by integrating monitoring, logging, and feedback mechanisms, while also enabling users to reenter the training and steering process even after deployment.

5 Platform Architecture and Implementation

Based on the design described in the previous section, we implemented a modular architecture to support the development of trustworthy AI models. The source code for our TrustAI platform can be found on GitHub.[11] The system architecture (cf. Fig. 2) builds upon multiple open-source libraries and frameworks and connects and extends them with custom components to create an efficient, extensible, and scalable training and deployment pipeline. The architecture is designed to operate in iterative cycles of training, evaluation, and refinement, allowing ML models to improve continuously based on user feedback.

At the configuration stage, AI experts define datasets, modeling objectives, and feature specifications through a structured manual setup. These settings are consolidated into a main configuration file, which acts as a declarative blueprint for task initialization. A dedicated service script parses these configurations to initialize downstream components (such as preprocessing workers and model trainers). The system then transitions into a preprocessing stage, where raw input data is transformed into standardized formats aligned with the task schema.

---

[11] https://github.com/fhstp/trustai-platform/, last visited 07/2025.

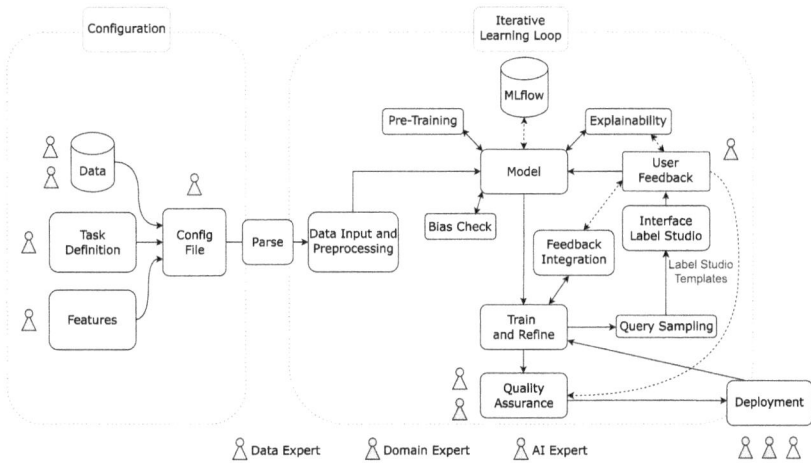

Fig. 2. System architecture of the TrustAI platform. Data, task definition, and features are specified by a domain and data expert and formalized into a config file by an AI expert (together with all parameters for the interactive model training). Next, the config file is parsed and an iterative learning process is established. The model is first pre-trained and then iteratively fine-tuned through user feedback at the level of labels and explanations. Once a certain quality is achieved the model is deployed.

The iterative learning loop is a central feature of the platform. A query sampling module selects data samples that are uncertain, misclassified, or belong to underrepresented or sensitive groups, incorporating various common sampling strategies from active learning. These examples are prioritized for annotation and presented to users via Label Studio. Once annotated (i. e. labels and explainability-driven ground truth), the feedback integration module processes the newly annotated data and provides it to the preprocessing and training components for iterative improvement of the model. Additionally, quality assurance services monitor model performance and fairness metrics across iterations, enforcing strict criteria before deployment of new model versions.

The platform is built on a Docker-based[12] infrastructure, with each module deployed as a container within an isolated, internally managed Docker network. Services communicate through defined APIs, supporting modular integration and future enhancements. At the core of the platform lies a model management and training module built around MLflow,[13] which systematically documents experiments, performance metrics, explainability outputs, and model artifacts, which are versioned and searchable to meet audit requirements and track performance trends. Model training is integrated into a feedback-driven pipeline, which includes a bias detection module that incorporates dedicated libraries like

---

[12] https://docker.com/, last visited 07/2025.
[13] https://mlflow.org/, last visited 07/2025.

AI Fairness 360[14] and an explainability module that integrates common methods such as Grad-CAM [24] and SHAP [16] for different ML models to provide decision explanations. Internally, these outputs are used for system diagnostics and quality assurance. Externally, explainability results serve as a communication channel to users, helping them first understand model predictions and then interact with them by providing explainability-related ground truth in the annotation interface, as part of the extended IML workflow. Label Studio serves as the backbone for annotation and feedback integration. We extend Label Studio through custom backend services that interface with MLflow and the training pipeline via the Label Studio API.[15] This tight coupling allows us to programmatically create annotation projects, synchronize model predictions with labeling tasks, and manage the ingestion of human feedback into the learning loop. To foster usability, we deploy Label Studio Templates that present model predictions and explanations directly within the annotation UI, enabling annotators to make informed corrections based on input data, predictions, and model behavior.

6 Outlook and Conclusion

The TrustAI platform addresses two key challenges in ML practice, the need for large labeled datasets and the lack of model transparency. By combining XAI with IML, the platform enables users to provide continuous feedback on model predictions and explanations, thereby steering the training process, improving performance, reducing bias, and aligning models with human reasoning. As a work in progress, this paper does not include an empirical evaluation of the platform and currently does not address privacy or security requirements. Future evaluations of the TrustAI platform in real-world applications, such as in clinical gait analysis, precision farming and cultural heritage, will demonstrate the adaptability and scalability of the platform in different domains. Its open-source nature and user-centered design democratize access to complex ML workflows while ensuring model transparency.

Acknowledgments. This research was funded by the Austrian Research Promotion Agency through the projects FO999904624 "FAIR-AI" and 898085 "TrustAI". The financial support by the Austrian Research Promotion Agency is gratefully acknowledged.

Disclosure of Interests. The authors have no competing interests to declare.

---

[14] https://github.com/Trusted-AI/AIF360, last visited 07/2025.
[15] https://labelstud.io/api, last visited 07/2025.

A Platform Criteria

To rigorously assess the suitability of the various available platforms (as described in Sect. 2), we defined several evaluation criteria capturing essential technical, usability, and ethical dimensions. These were:

1. **Licensing:** openness and permissiveness of the software license.
2. **Data privacy and protection:** support for local deployment and data governance.
3. **Backend language:** compatibility with Python or other accessible languages.
4. **Documentation:** availability and comprehensiveness of guides and API references.
5. **Code quality:** modularity, readability, and structure of the backend.
6. **Scalability:** support for large-scale data or user loads.
7. **Community support:** active user base, discussion forums, and responsiveness.
8. **Integration ease:** ability to interface with ML frameworks and pipelines.
9. **Security:** presence of authentication, encryption, or role-based access.
10. **Performance:** runtime efficiency and responsiveness.
11. **Active development:** frequency of updates and codebase activity.
12. **Framework compatibility**: interoperability with tools like TensorFlow, PyTorch, etc.
13. **UI customizability:** ability to modify or extend the user interface.
14. **UI component library:** availability of reusable and modular components.
15. **Deployment simplicity:** ease of installation and running in local/cloud environments.
16. **Data source interoperability:** flexibility in connecting to diverse data backends.
17. **Format support:** compatibility with various data types and annotation schemas.
18. **Model versioning and experiment tracking:** built-in or pluggable tracking features.
19. **Community feedback:** qualitative reviews and third-party endorsements.
20. **Accessibility and inclusivity:** attention to UI accessibility and global usability.

B Platform Evaluation

Each platform was scored on these twenty criteria on a scale from 1 to 5, where 1 indicated full compliance with project needs and 5 signaled critical incompatibilities. Any platform receiving a score of 5 in any category was disqualified, as such scores reflected fundamental limitations incompatible with our project's requirements.

Table 1. Summary table of scores for the eleven platforms examined against the twenty criteria described in Sect. A.

Criterion	BigML	InterprML	Labelbox	Labelling	Label Studio	Marcelle	Neural Network Console	PerceptiLabs	Prodigy 101	Supervisely	TornadoAI
Licensing	5	1	5	1	1	1	5	×	5	5	1
Data privacy and protection	×	1	×	×	1	1	×	×	1	×	1
Backend language	×	1	×	×	1	1	×	×	1	×	2
Documentation	×	2	×	×	1	2	×	×	1	×	2
Code quality	×	1	×	×	1	1	×	×	2	×	2
Scalability	×	1	×	×	1	2	×	×	5	×	3
Community support	×	1	×	×	1	2	×	×	2	×	5
Integration ease	×	2	×	×	1	2	×	×	2	×	3
Security	×	2	×	×	2	2	×	×	2	×	×
Performance	×	2	×	×	2	3	×	×	3	×	3
Active development	×	2	×	5	1	4	×	×	1	×	5
Framework compatibility	×	4	×	×	1	3	×	×	2	×	2
UI customizability	×	5	×	×	1	2	×	5	2	×	2
UI component library	×	5	×	×	1	2	×	5	1	×	4
Deployment simplicity	×	2	×	×	1	2	×	×	1	×	3
Data source interoperability	×	4	×	×	1	3	×	×	2	×	2
Format support	×	3	×	×	1	3	×	×	2	×	3
Model versioning and experiment tracking	×	4	×	×	2	4	×	×	3	×	3
Community feedback	×	2	×	5	1	×	×	×	2	×	5
Accessibility and inclusivity	×	3	×	×	2	2	×	×	2	×	5

Abbreviations: Rating scale from 1 ("fully supported or excellent") to 5 ("not supported"); × means "no evaluation" (because of other essential criteria whose scores already disqualify the platform as a viable option).

C Platform Requirements

The full set of requirements derived from the workshop (as described in Sect. 3) was as follows:

Data Processing

D1 The system is GDPR-compliant.
D2 The system works with anonymized data.
D3 The system works without sharing sensitive data (i. e. no cloud solutions).

Human-Centered Design

H1 The system provides specific solutions for different user groups: AI experts, domain experts, generic users.
H2 The system ensures user acceptance through suitable descriptions.
H3 The system allows for different user backgrounds (e. g. language skills).
H4 The system can provide personalized models (e. g. specific to a single user).

Interactivity

I1 The system and its users learn from each other.
I2 The system requires as few annotations as possible.

I3 The system needs user input for complex or unusual data.

Flexibility

F1 The system avoids vendor lock-in.
F2 The system supports both software-as-a-service and on-premises use.
F3 The system scales appropriately with increasing volume of data or customers.
F4 The system supports use cases with little data, big data and bad data quality.

Transparency/Explainability

T1 The system supports transparency and XAI for all user groups.
T2 The system enhances explainability through visualization.
T3 The system helps with recognition of the causes of anomalies (and not just with anomaly detection).
T4 The system supports comparing results of different models and recognizing different models' strengths.
T5 The system facilitates outputting confidence scores.

References

1. Adadi, A., Berrada, M.: Peeking inside the black-box: a survey on explainable artificial intelligence (XAI). IEEE Access **6**, 52138–52160 (2018). https://doi.org/10.1109/ACCESS.2018.2870052
2. Amershi, S., et al.: Guidelines for human-AI interaction. In: Proceedings of the 2019 CHI Conference on Human Factors in Computing Systems CHI 2019, ACM, New York (2019). https://doi.org/10.1145/3290605.3300233
3. Bach, S., Binder, A., Montavon, G., Klauschen, F., Müller, K.R., Samek, W.: On pixel-wise explanations for non-linear classifier decisions by layer-wise relevance propagation. PLoS One **10**(7) (2015). https://doi.org/10.1371/journal.pone.0130140
4. Bansal, G.: Explanatory dialogs: towards actionable, interactive explanations. In: Proceedings of the 2018 AAAI/ACM Conference on AI, Ethics, and Society, AIES 2018, pp. 356–357. ACM, New York (2018). https://doi.org/10.1145/3278721.3278795
5. Chromik, M., Butz, A.: Human-XAI interaction: a review and design principles for explanation user interfaces. In: Human-Computer Interaction – INTERACT 2021, pp. 619–640. LNCS, vol. 12933, Springer, Cham (2021). https://doi.org/10.1007/978-3-030-85616-8_36
6. Dudley, J.J., Kristensson, P.O.: A review of user interface design for interactive machine learning. ACM Trans. Interact. Intell. Syst. **8**(2) (2018). https://doi.org/10.1145/3185517
7. Fogarty, J., Tan, D., Kapoor, A., Winder, S.: CueFlik: interactive concept learning in image search. In: Proceedings of the 2008 SIGCHI Conference on Human Factors in Computing Systems, CHI 2008, pp. 29–38. ACM, New York (2008). https://doi.org/10.1145/1357054.1357061
8. Friedman, B., Nissenbaum, H.: Bias in computer systems. ACM Trans. Inf. Syst. **14**(3), 330–347 (1996). https://doi.org/10.1145/230538.230561

9. Gao, Y., Gu, S., Jiang, J., Hong, S.R., Yu, D., Zhao, L.: Going beyond XAI: a systematic survey for explanation-guided learning. ACM Comput. Surv. **56**(7) (2024). https://doi.org/10.1145/3644073
10. Gao, Y., Sun, T.S., Zhao, L., Hong, S.R.: Aligning eyes between humans and deep neural network through interactive attention alignment. Proc. ACM Hum.-Comput. Interact. **6**(CSCW2) (2022). https://doi.org/10.1145/3555590
11. Kapania, S., Taylor, A.S., Wang, D.: A hunt for the snark: annotator diversity in data practices. In: Proceedings of the 2023 CHI Conference on Human Factors in Computing Systems, CHI 2023. ACM, New York (2023). https://doi.org/10.1145/3544548.3580645
12. Kim, S.S.Y., Watkins, E.A., Russakovsky, O., Fong, R., Monroy-Hernández, A.: "Help me help the AI": understanding how explainability can support human-AI Interaction. In: Proceedings of the 2023 SIGCHI Conference on Human Factors in Computing Systems, CHI 2023. ACM, New York (2023). https://doi.org/10.1145/3544548.3581001
13. Liang, P., Song, B., Zhan, X., Chen, Z., Yuan, J.: Automating the training and deployment of models in MLOps by integrating systems with machine learning. In: Proceedings of the 2nd International Conference on Software Engineering and Machine Learning, CONF-SEML 2024, Applied and Computational Engineering 76, EWA Publishing, Oxford (2024). https://doi.org/10.54254/2755-2721/76/20240690
14. Long, D., Magerko, B.: What is AI literacy? Competencies and design considerations. In: Proceedings of the 2020 CHI Conference on Human Factors in Computing Systems, CHI 2020, ACM, New York (2020). https://doi.org/10.1145/3313831.3376727
15. Lu, Q., Zhu, L., Xu, X., Whittle, J., Zowghi, D., Jacquet, A.: Responsible AI pattern catalogue: a collection of best practices for AI governance and engineering. ACM Comput. Surv. **56**(7) (2024). https://doi.org/10.1145/3626234
16. Lundberg, S.M., Lee, S.I.: A unified approach to interpreting model predictions. In: Proceedings of the 31st International Conference on Neural Information Processing Systems, NIPS 2017, pp. 4768–4777. Curran Associates, Red Hook, NY (2017). http://papers.nips.cc/paper/7062-a-unified-approach-to-interpreting-model-predictions.pdf
17. Mathewson, K.W.: A human-centered approach to interactive machine learning. In: Proceedings of the 4th Multi-Disciplinary Conference on Reinforcement Learning and Decision Making, pp. 460–464. RLDM 2019 (2019)
18. Mehrabi, N., Morstatter, F., Saxena, N., Lerman, K., Galstyan, A.: A survey on bias and fairness in machine learning. ACM Comput. Surv. **54**(6) (2021). https://doi.org/10.1145/3457607
19. Phillips, P.J., et al.: Four principles of explainable artificial intelligence (2021). https://doi.org/10.6028/NIST.IR.8312
20. Ross, A.S., Hughes, M.C., Doshi-Velez, F.: Right for the right reasons: training differentiable models by constraining their explanations. In: Proceedings of the Twenty-Sixth International Joint Conference on Artificial Intelligence, IJCAI 2017, pp. 2662–2670. IJCAI Organization, California (2017). https://doi.org/10.24963/ijcai.2017/371
21. Safdar, N.M., Banja, J.D., Meltzer, C.C.: Ethical considerations in artificial intelligence. Eur. J. Radiol. **122** (2020). https://doi.org/10.1016/j.ejrad.2019.108768
22. Scheutz, M.: Ethical aspects and challenges for interactive task learning. In: Gluck, K.A., Laird, J.E. (eds.) Interactive Task Learning: Humans, Robots, and Agents

Acquiring New Tasks through Natural Interactions, pp. 295–304. MIT Press, Cambridge, MA (2019)
23. Schöning, J., Kruse, N.: Compliance of AI systems (2025). https://doi.org/10.48550/arXiv.2503.05571
24. Selvaraju, R.R., Cogswell, M., Das, A., Vedantam, R., Parikh, D., Batra, D.: Grad-CAM: visual explanations from deep networks via gradient-based localization. In: 2017 IEEE International Conference on Computer Vision, ICCV 2017, pp. 618–626. IEEE, Washington, DC (2017). https://doi.org/10.1109/ICCV.2017.74
25. Shneiderman, B.: Human-Centered AI. Oxford University Press, Oxford (2022). https://doi.org/10.1093/oso/9780192845290.001.0001
26. Teso, S., Alkan, Ö., Stammer, W., Daly, E.: Leveraging explanations in interactive machine learning: an overview. Front. Artif. Intell. **6** (2023). https://doi.org/10.3389/frai.2023.1066049
27. Teso, S., Kersting, K.: Explanatory interactive machine learning. In: Proceedings of the 2019 AAAI/ACM Conference on AI, Ethics, and Society, AIES 2019, pp. 239–245. ACM, New York (2019). https://doi.org/10.1145/3306618.3314293
28. Thelwall, M.: Gender bias in machine learning for sentiment analysis. Online Inf. Rev. **42**(3), 343–354 (2018). https://doi.org/10.1108/OIR-05-2017-0153
29. Thomaz, A.L., et al.: Interaction for task instruction and learning. In: Gluck, K.A., Laird, J.E. (eds.) Interactive Task Learning: Humans, Robots, and Agents Acquiring New Tasks through Natural Interactions, pp. 91–110. MIT Press, Cambridge, MA (2019)
30. Vallor, S.: The AI Mirror: How to Reclaim Our Humanity in an Age of Machine Thinking. Oxford University Press, Oxford (2024). https://doi.org/10.1093/oso/9780197759066.001.0001

Collaborative Trustworthy Foundation Model Framework: An Environmental Sustainability Use-Case to Detect Contamination Objects in Organic Waste Streams

Alexander Valentinitsch[1](✉), Batuhan Bencik[1], Mathias Brucker[1], Gregor Lammer[2], Cornelia Adami[2], Mohit Kumar[1], Lukas Fischer[1], and Florian Kromp[1]

[1] Software Competence Center Hagenberg GmbH, Softwarepark 32a, Hagenberg 4232, Austria
`alexander.valentinitsch@scch.at`
[2] Brantner Digital Solutions GmbH, Dr. Franz Wilhelmstraße 2a, 3500 Krems an der Donau, Austria

Abstract. This work introduces a privacy-preserving framework that integrates foundation models with federated learning through a synergistic application using Low-Rank Adaptation (LoRA). Our three-stage pipeline—centralized pretraining, federated fine-tuning, and knowledge distillation—enables efficient and GDPR-compliant model updates. Using Florence-2 and the Hugging Face PEFT library, we demonstrate the framework on contamination object detection in organic waste streams across five distributed sites. Results show that the federated fusion model outperforms both centralized and local baselines in terms of IoU and detection accuracy, highlighting the effectiveness of LoRA-based adaptation for real-world, decentralized settings..

Keywords: Trustworthy AI · Foundation Models · Federated Learning · Low-Rank Adaptation · Vision-Language · Organic Waste · Environmental Sustainability

1 Introduction

Vision-language models (VLMs) are increasingly recognized as foundation models, particularly when trained on extensive and diverse datasets, enabling robust

---

The research reported in this paper has been supported by the AI Mission Austria Flagship Project FAIR-AI [(FFG grant no. 904624)] and BMIMI, BMWET, and the State of Upper Austria in the frame of the SCCH competence center INTEGRATE [(FFG grant no. 892418)] part of the FFG COMET Competence Centers for Excellent Technologies Programme.

generalization across multiple downstream tasks without extensive task-specific fine-tuning [2]. Foundation models demonstrate the ability to transfer generalized knowledge acquired from massive multi-modal datasets—including images and text—to various tasks such as image captioning, visual question answering, and object detection, often requiring no or only minimal additional supervised training for new applications [32]. Leveraging these capabilities, this study explores a real-world use case in contamination detection within organic waste streams. Effective organic waste sorting critically depends on accurately identifying contaminants, including plastics, metals, and glass. By framing contamination detection as a multi-label object detection problem, we can later calculate a contamination score representing the ratio of detected contaminants relative to total objects per image. This practical scenario underscores both the environmental significance and the real-world applicability of federated fine-tuning strategies for foundation models. At the same time, privacy concerns are highly relevant in this domain: image data collected from municipal waste bins may inadvertently reveal personal or household-level information. Ensuring compliance with data protection regulations like the GDPR is therefore essential, especially in public-sector deployments involving citizen-facing infrastructure.

To effectively adapt such large-scale foundation models in decentralized, privacy-sensitive scenarios, parameter-efficient fine-tuning techniques are essential. In this context, Low-Rank Adaptation (LoRA) provides a particularly promising solution, inserting lightweight, trainable matrices into selected layers of the frozen pretrained model [10]. These LoRA adapters can efficiently be trained locally for each federated client and subsequently aggregated at a central server site, significantly reducing communication overhead and enabling privacy-preserving model refinement [1, 9, 13, 24, 30, 31]. Florence-2, introduced by Li et al. [28], offers a unified vision-language architecture that is well-suited for adaptation in federated settings. Its ability to generalize across a wide range of tasks makes it a strong foundation model for decentralized applications.

In this work, we leverage Florence-2 as the base model and adapt it using Low-Rank Adaptation (LoRA) modules trained via the Hugging Face's Parameter-Efficient Fine-Tuning (PEFT) library [16]. As demonstrated by our results this approach enables efficient and privacy-preserving fine-tuning on client-specific data, aligning with the constraints of non-IID data distributions and communication efficiency in federated learning environments.

2 Related Work

Foundation Models. Florence-2 [28], is a state-of-the-art vision-language foundation model designed to handle a wide spectrum of visual tasks through a unified, prompt-based architecture. It builds on a sequence-to-sequence design and is trained on the FLD-5B dataset, which contains over 5.4 billion annotations across 126 million images. This massive and diverse dataset allows Florence-2 to effectively generalize across tasks such as object detection, segmentation, captioning, visual grounding, OCR, and dense region understanding, both in zero-shot and fine-tuned configurations.

Unlike earlier models that rely on task-specific heads (e.g., Faster R-CNN [21]) or contrastive pretraining objectives (e.g., CLIP [19]), Florence-2 employs a uniform training strategy using textual prompts and a shared loss function across tasks. This enables seamless task switching and fine-tuning, making it particularly well-suited for federated learning settings where site-specific data may be scarce or task-specific annotations may not be uniformly available. Its design enables rich semantic reasoning across varying levels of visual hierarchy and granularity, which is crucial for tasks like waste contamination detection that require both local object recognition and accurate classification of objects by their semantic category.

In comparison to other vision-language models, Florence-2 offers distinct advantages. CLIP [19], for instance, is optimized for zero-shot classification and retrieval through contrastive learning, but lacks generative and multitask capabilities. While extensions such as WinCLIP [7] improve CLIP's performance on fine-grained visual recognition through hierarchical attention mechanisms, and AnomalyCLIP [11] adapts it for zero-shot anomaly detection and localization, these models remain specialized and do not offer the broader task flexibility and prompt-driven generalization available in foundation models like Florence-2. Kosmos-2 [18] and LLaVA [15] extend vision-language integration through grounding and instruction-following, but depend on large language models or domain-specific tuning, which can limit their general-purpose applicability. YOLO [20] is highly efficient for real-time object detection and its recent extension, YOLO-World [3], introduces open-vocabulary detection by integrating vision-language features.

Florence-2 outperforms many of these models in both zero-shot and fine-tuned settings, despite having fewer parameters (0.77B for Florence-2-large vs. 1.6B for Kosmos-2) [28]. Experimental evidence indicates that Florence-2, when fine-tuned appropriately, achieves object detection performance comparable to specialized models such as YOLOv8 in unstructured environments [22]. Its modular design and scalable variants also make it adaptable to different resource constraints, positioning it as a versatile choice for federated and sustainability-focused deployments.

Federated Learning. The classical approach to federated learning, introduced by McMahan et al. (2017) [17], aims to train a shared global model through the iterative process of (1) training local models on client data and (2) aggregating the resulting model updates to refine a central global model. However, because client data is sampled from distinct and often non-identical distributions, designing and analyzing efficient federated learning algorithms remains a major challenge [29]. Key obstacles include handling data heterogeneity, ensuring communication efficiency, and maintaining computational scalability. Strategies for addressing heterogeneity include personalized model learning for each client [4], model-agnostic meta-learning frameworks [5], and selective parameter sharing across clients [25]. Other approaches such as adversarial federated learning [14] and clustered federated learning (CFL) [23] have also been developed to manage diverse data distributions and group clients based on similarity. While

CFL can be communication-intensive, its cost may be mitigated by leveraging data distribution signatures at the server to initialize optimal client clusters [27], which also offers a seamless mechanism for integrating new participants into the federation.

The intersection of federated learning and foundation models has opened new research directions in AI [32]. Federated learning supports foundation model development by enabling privacy-preserving training on distributed data sources [6], while foundation models contribute pretrained general knowledge that accelerates convergence and enhances performance in heterogeneous environments. To adapt foundation models efficiently, parameter-efficient fine-tuning methods like Low-Rank Adaptation (LoRA) have been proposed [10]. These approaches typically freeze the foundation model's parameters and update only a small adapter module. Recent research demonstrates how LoRA can be used in federated setups to enable scalable and communication-efficient fine-tuning [24]. By federating only the lightweight adapters, communication overhead is drastically reduced and can be further optimized using adaptive and sparse update mechanisms [13]. To address intra-client variability, personalized FL methods employing dual LoRA modules have been introduced to jointly capture global and client-specific representations [30].

Knowledge Fusion. Knowledge distillation and model fusion have emerged as pivotal strategies in enhancing the scalability and adaptability of vision-language foundation models. Knowledge distillation refers to the process of transferring knowledge from a larger, often more accurate teacher model to a smaller, more efficient student model, typically by aligning output logits, intermediate features, or relational structures [8]. This enables significant reductions in computational cost and memory usage while maintaining comparable performance—an essential property for deployment in federated environments. Model fusion complements distillation by integrating outputs or learned representations from multiple specialized models, thereby improving overall robustness and task generalization. This is particularly effective in federated contexts, where clients may adapt to different data distributions, and fusion can help to consolidate those domain-specific learnings into a unified model.

3 Methodology

Our methodology is structured as a three-stage cycle that continuously evolves the foundation model through a combination of centralized pretraining and decentralized collaborative learning (as visualized in Fig. 1):

Stage 1: Centralised Pretraining and Model Development. At this stage, we leverage diverse and anonymized data sources including text annotation and images, to enhance the existing robust foundation model with centralized pretraining. Florence-2 serves as our base architecture, which is further finetuned through LoRA adapters.

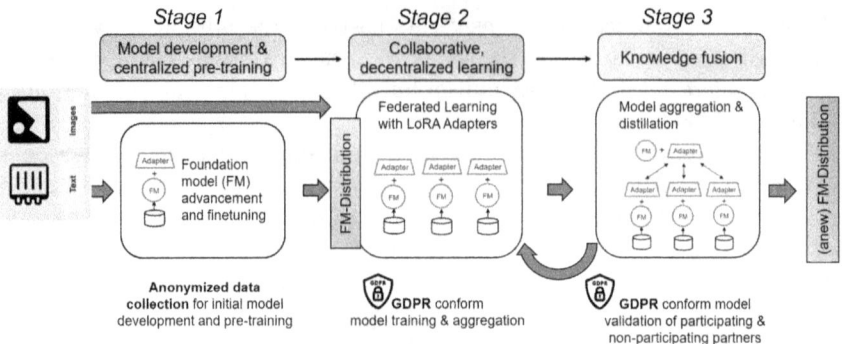

Fig. 1. Overview of the proposed three-stage pipeline. Stage 1 involves centralized pretraining on anonymized, multi-source data. In Stage 2, the pretrained model is distributed to clients and fine-tuned locally using LoRA adapters in a federated learning setup. Stage 3 aggregates the LoRA adapters, applies optional distillation, and initializes a new cycle with an updated foundation model.

Stage 2: Federated Fine-Tuning with LoRA Adapters. Our approach is inspired by FedLoRA, a model-heterogeneous federated learning framework proposed by [30], which introduces the use of homogeneous low-rank adapters to reduce communication and computational costs. In our setting, clients utilize the Hugging Face's PEFT library to inject and fine-tune LoRA adapters into the Florence-2 base model. We adopt its efficient and scalable strategy of sharing and aggregating LoRA updates. This enables federated model refinement while maintaining GDPR compliance, as all training occurs locally and only adapter parameters are exchanged.

Stage 3: Model Aggregation and Knowledge Distillation. Adapted LoRA parameters from clients are aggregated at a central server, followed by model distillation techniques to fuse the collective knowledge into a new, evolved foundation model. We adopt a modular fusion strategy using Hugging Face's PEFT library [16] to aggregate client-specific LoRA adapters and merge them back into the shared Florence-2 base model. This operation acts as a lightweight form of distillation, allowing the central model to integrate task-adapted knowledge without retraining the entire network. Together, these techniques enable scalable, communication-efficient adaptation of foundation models across heterogeneous client environments, enhancing both their real-world usability and environmental sustainability. This distilled model undergoes validation for generalizability and compliance before re-entering Stage 2, initiating another refinement cycle.

This cyclical approach supports iterative improvement of the distributed foundation model while enabling real-world deployment scenarios such as contamination detection in organic waste.

Detect Contamination Objects. To validate this framework in a real-world context, we introduce a use case focused on contamination object detection in organic waste streams. This task is essential for calculating a contamination score, which is defined as the ratio of detected contaminant objects to the total number of detected objects in a given image. Monitoring contamination levels in municipal waste bins has become increasingly important for improving waste sorting efficiency and meeting recycling targets.

The dataset used in this study comprises images collected from multiple geographic regions, each annotated with object-level labels. These labels span both organic materials, such as food scraps, and contaminants, including plastics, metals, glass, and cardboard. However, the distribution of these labels is highly imbalanced: a small number of classes, such as plastic foil and paper handkerchiefs, dominate the dataset, while many other contamination types appear infrequently. This long-tail distribution introduces challenges for both model training and contamination scoring, particularly in sites where rare contaminants are more prevalent. As a result, accurate object detection is not only critical for classification performance but also directly impacts the reliability of the contamination score used for downstream waste assessment and decision-making.

Each client in our FL setup corresponds to a distinct region with unique waste characteristics. The Florence 2 model, fine-tuned using LoRA at each client site, learns both shared and localized visual patterns of contamination. The federated LoRA parameters are then aggregated to build a generalized model capable of accurate contamination assessment across regions. This setup ensures data privacy, and demonstrates the adaptability of our system to sustainability-related applications.

4 Experiments

Datasets. Our experiments are conducted on a custom-curated dataset collected from five geographically distributed waste management sites (labeled A to E). Each site contributed a distinct number of raw training and test samples, reflecting the variability in regional waste compositions and collection practices. A sixth site, referred to as Validation, was held out exclusively for validation purposes to assess generalization. An overview of the raw and augmented Image counts per site can be found in Table 1.

To prepare the data for Florence-2's object detection pipeline, we applied object-centric augmentation. This ensured that the resulting inputs retained sufficient context while remaining compatible with Florence-2's architectural input requirements.

Label Harmonization. The raw dataset initially contained 34 highly granular and inconsistently named object classes. Several labels were semantically overlapping or represented variations of the same object (e.g., `plastic_foil`, `plastic_foil_bag`, `plastic_bag_color`), which introduced noise and inconsistencies during training. To address this, we developed a rule-based label mapping strategy, grouping similar object classes under harmonized, semantically meaningful categories.

Table 1. Raw and Augmented Image Counts per Site

Site	Raw Train	Raw Test	Aug. Train	Aug. Test
A	587	147	3286	837
B	835	209	5417	1404
C	825	207	2294	640
D	2343	586	13599	3175
E	130	33	1008	278
Validation	160	40	1032	243

The label harmonization process reduced the original 34 classes to 17 semantically coherent categories that better represent functional roles in contamination assessment. Figure 2 visualizes the label distribution after harmonization.

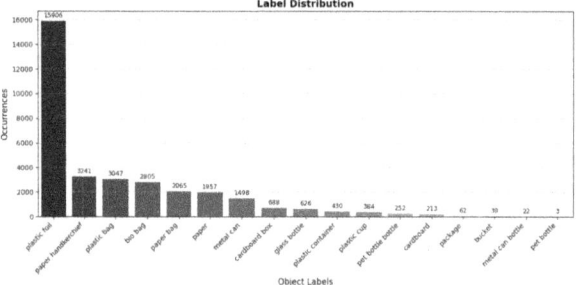

Fig. 2. Data distribution after label harmonization. It illustrates the post-mapping distribution over 17 unified labels used for training. The harmonization step reduces redundancy and improves consistency across decentralized datasets.

Evaluation Metrics. To evaluate model performance in the object detection task, we used two complementary metrics: Intersection over Union (IoU) and the confusion matrix. IoU quantifies the spatial overlap between a predicted (i.e. Inference (Inf) bounding box and the corresponding ground truth (GT) box. This metric ranges from 0 to 1, with higher values indicating better localization accuracy. IoU is especially relevant for tasks where precise object boundaries affect downstream interpretation, such as contamination detection in waste images. In addition, we show a confusion matrix to assess classification performance across the defined object categories.

Experimental Baseline. To evaluate the effectiveness of our federated adaptation approach using Florence-2 with LoRA, we designed a two-part experimental baseline that captures both global generalization and site-specific performance.

In the first part of the evaluation, we test the final global model—obtained after aggregating LoRA parameters from all participating sites—on the test set of each individual site (A–E). This scenario reflects a common deployment context in federated learning, where a shared global model must generalize across clients with diverse data distributions. By evaluating the same model across all sites, we assess how well the collaborative training process captures the underlying variability in contamination patterns. As can be observed from Table 2A, the global model outperforms all local models by a high margin, irrespective of the site applied. Despite this high performance in object detection, the confusion matrices (Fig. 3) outline that the global model struggles to robustly learn correct class-assignment of the objects detected.

In the second part of the evaluation, we compare the performance of the plain Florence-2 model, the locally fine-tuned (BASE) model and the aggregated, globally distilled model (FUSION). As demonstrated in Table 2B, the FUSION model outperforms the BASE model and the plain Florence-2 model in terms of IoU. However, the class-assignment of detected objects remains limited (see (Fig. 4)), irrespective of the model.

Table 2. Comparison of Intersection over Union (IoU) across different model configurations. **(A)** Local vs. Global: Local models are Florence-2 instances fine-tuned with LoRA using data from a single site only, without any form of data sharing. Global models are fine-tuned using aggregated LoRA updates across all sites in a federated learning setup. **(B)** BASE vs. FUSION: The BASE model is fine-tuned centrally on pooled training data using LoRA. The FUSION model is globally distilled by aggregating LoRA adapters from all federated clients and merging them with the BASE backbone. The plain Florence-2 model ("microsoft/Florence-2-base") [28].

Site	Local IoU	Global IoU
A	0.045	0.612
B	0.018	0.545
C	0.058	0.577
D	0.015	0.590
E	0.393	0.590

A

Model	Mean IoU
Florence-2	0.392
BASE	0.563
FUSION	0.679

B

Federated Training Setup. Federated training was conducted across five client sites (A–E), each holding a unique subset of the training data as outlined earlier. The training process followed a round-based approach using the Federated Averaging (FedAvg) algorithm to aggregate LoRA adapter updates from the local clients. During each communication round, the server distributed the frozen Florence-2 base model along with the current global LoRA parameters to all clients.

Each client performed local training exclusively on its own dataset, updating only the LoRA-injected layers while keeping the backbone model frozen. Local training was conducted using stochastic gradient descent (SGD) with a learning rate of 0.001, a batch size of 32, and for 3 epochs per round. All clients participated in each round (i.e., client fraction $C = 1.0$), and the training ran for a total of 10 communication rounds.

After each round, the LoRA parameters were sent back to the server, where they were aggregated using weighted averaging proportional to the number of training samples at each site. The aggregated adapter was then broadcast to all clients for the next round. This process continued until convergence. No client shared raw data or model activations, ensuring full compliance with GDPR-aligned privacy constraints.

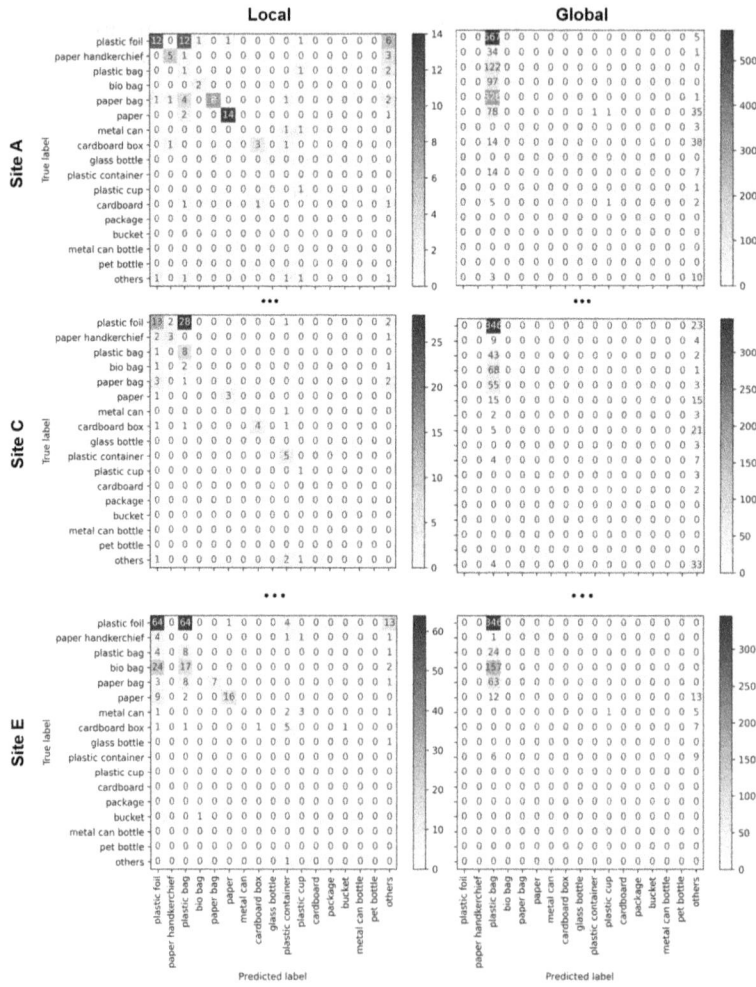

Fig. 3. Confusion matrices of local vs. global models for Sites A, C, and E. These plots reveal substantial inter-class confusion despite good localization performance, motivating future improvements in classification robustness.

Framework Evaluation. To evaluate the performance of our end-to-end pipeline, we conducted an additional experiment focused on model development and fusion. In Stage 1, we fine-tuned the publicly available Florence-2-base model ("microsoft/Florence-2-base") [28] using a centralized approach, where the training datasets from all five sites (A–E) were combined. Rather than performing full fine-tuning, we applied LoRA-based adaptation to preserve the pretrained model's generalization capabilities and reduce the risk of overfitting. After fine-tuning, we merged the LoRA adapters into the backbone using the PEFT library's, producing a task-specific central BASE model. This BASE model was then distributed to each site as an initialization for Stage 2, where federated fine-tuning continued on the remaining test sets from each site. In Stage 3, the resulting LoRA adapters from the federated clients were aggregated using FedAvg and fused back with the BASE model to form a final FUSION model.

We compared the performance of the Florence-2-base model, the centralized BASE model, and the final FUSION model on the complete validation set (concatenated train and test). Results are shown in Table 2B and few samples are visualized in Fig. 5.

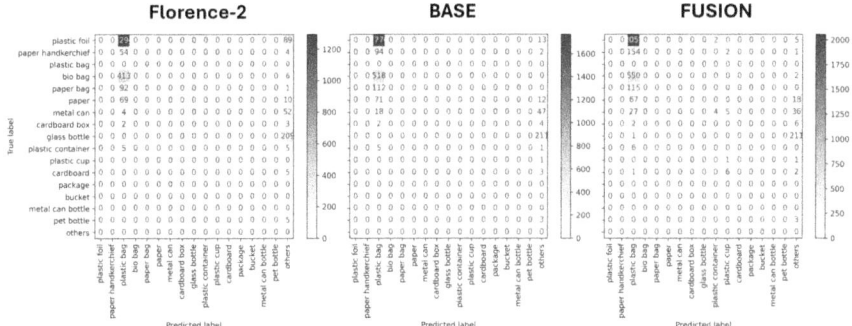

Fig. 4. Confusion matrices of the centralized LoRA fine-tuned base model (BASE) and the federated fusion model (FUSION) in comparison with the Florence-2 model ("microsoft/Florence-2-base") [28].

5 Discussion

The results presented in Table 2 show a consistent trend: the global model significantly outperforms the locally fine-tuned models across nearly all sites in terms of mean Intersection over Union (IoU). Local models in this context refer to Florence-2 instances fine-tuned with LoRA adapters using only the data available at each individual site, without any federated aggregation or knowledge sharing. For example, at Site D, the global model achieves an IoU of 0.590, while the

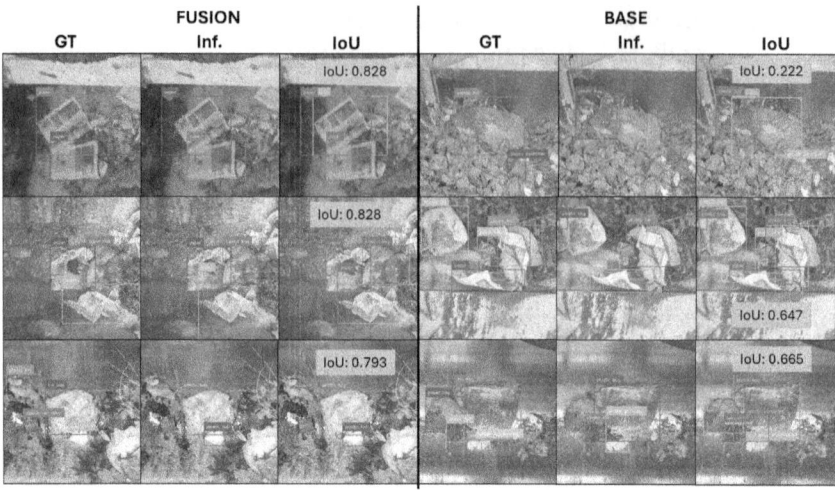

Fig. 5. Visualized inference (Inf.) results from the BASE and FUSION models against ground truth (GT) annotations, including computed Intersection over Union (IoU) scores.

local model reports just 0.015. Similar gaps are observed at Sites A through C, highlighting the robustness of the globally pre-trained model. Florence-2 misassigns classes, particularly collapsing many predictions into dominant categories like "plastic bag" or "plastic foil." This suggests that the model captures spatial presence effectively in a federated setup, but struggles to learn robust interclass boundaries (Fig. 3). This is most likely use to the long-tailed and imbalanced label distribution across clients. (see Figure 2) and could be mitigated by strategies such as class-balanced sampling or balanced loss weighting. If the misalignment is based on a domain shift between images of different sites, domain adaptation strategies could be applied. The adaptation of the model with respect to correct class assignment with high performance is left for future work.

The BASE model, created through centralized LoRA fine-tuning, begins to recover minority classes such as "glass bottle" and "pet bottle" but still leans heavily on "plastic bag." The FUSION model, derived from federated aggregation of LoRA adapters, shows the most balanced behavior—retaining high IoU (0.679) while improving classification across more diverse object types, including "metal can," "paper handkerchief," and "plastic cup."

These findings underscore a key trade-off: global models offer reliable bounding box alignment, while local models bring greater class specificity. Federated fusion leverages both, yielding improved spatial and semantic performance. This highlights the value of personalized federated learning approaches that combine centralized pretraining with personalized adaptation in decentralized contexts. A comparison of the fused model to a centrally trained model, trained on all data from all sites, would be valuable to assess the effect of globally aggregating local models. However, this is not possible in our setting since data is not allowed to be shared. Despite semantic misclassifications, the model remains practically useful due to its strong IoU performance, which ensures reliable object localization even when class labels are ambiguous. At the same time, this highlights the key advantage of the collaborative foundation model approach: enabling training on large, distributed datasets while preserving data privacy.

While the proposed framework effectively demonstrates federated fine-tuning with LoRA on Florence-2 for contamination detection, some limitations remain. The object label distribution was highly imbalanced across sites, with dominant classes such as plastic foil overshadowing underrepresented categories, limiting both global and local model generalization. Local models, particularly at Sites A–D, suffered from data scarcity and exhibited poor IoU despite greater label diversity, indicating underfitting and semantic drift. Moreover, the evaluation focused on final model performance without analyzing intermediate training rounds or varying aggregation strategies. Lastly, although LoRA reduces communication overhead, the framework lacks explicit mechanisms for personalization [26], which may be critical in settings with strong client heterogeneity.

To better address data heterogeneity, we plan to incorporate clustered federated learning (CFL) [23, 27]. Rather than training a single global model, CFL will allow clients to be grouped based on data similarity, enabling more specialized local adaptations. We will derive compact, privacy-preserving data signatures using geometrical-inspired kernel machines [12], allowing the server to cluster clients without accessing raw data. Each cluster would then train and adapt its own model variant using LoRA, improving personalization and performance, particularly for underrepresented sites. Additionally, improving class-level accuracy in federated models remains a challenge. Future work will also explore class-balanced sampling and reweighting strategies to reduce inter-class confusion under non-IID data distributions.

While our evaluation focuses on organic waste contamination, the proposed framework is broadly applicable to other domains with similar privacy constraints. In particular, medical imaging presents a compelling use case, where sensitive patient data is distributed across institutions. We plan to explore this generalization in future work.

Florence-2's built-in support for vision-language tasks, such as image captioning and scene-level descriptions using prompt tokens, provides an additional semantic layer for interpreting waste imagery. This functionality could be leveraged in future work to complement object detection with contextual understanding—potentially enabling contamination classification not just by detected items but also by inferred image semantics, such as waste setting, composition, or packaging context (Fig. 6).

Fig. 6. Visual analysis of a sample image. Left: object detection <OD> output including ground truth (GT) and model inference (Inf.). Right: semantic descriptions generated by Florence-2 using the <CAPTION> and <DETAILED_CAPTION> prompts. These captions offer complementary semantic information that could improve contextual understanding and contamination classification.

6 Conclusion

We presented a collaborative framework for privacy-preserving adaptation of foundation models across decentralized environments using federated learning and Low-Rank Adaptation (LoRA). By leveraging Florence-2 and Hugging Face's PEFT library, the approach enables efficient training with minimal communication overhead while maintaining data privacy. Applied to the task of contamination detection in organic waste streams, the framework achieved robust performance across diverse deployment contexts. Evaluations across five sites showed strong object localization performance from the federated fusion model, reflected in the highest mean IoU. However, confusion matrix analysis revealed persistent class-level misclassifications, underscoring that while spatial accuracy is robust in federated settings, semantic classification remains limited by data imbalance and heterogeneity. Future work will explore clustered federated learning guided by data distribution signatures to improve personalization and reduce semantic confusion, further enhancing the adaptability of foundation models in non-IID scenarios.

References

1. Babakniya, S., et al.: Slora: federated parameter efficient fine-tuning of language models. arXiv preprint arXiv:2308.06522 (2023)
2. Bommasani, R., et al.: On the opportunities and risks of foundation models. arXiv preprint arXiv:2108.07258 (2021)
3. Cheng, T., Song, L., Ge, Y., Liu, W., Wang, X., Shan, Y.: Yolo-world: Real-time open-vocabulary object detection. arXiv preprint arXiv:2401.17270 (2024)
4. Collins, L., Hassani, H., et al.: Exploiting shared representations for personalized federated learning. In: NeurIPS Workshop on Federated Learning for User Privacy and Data Confidentiality (2021)

5. Fallah, A., Mokhtari, A., Ozdaglar, A.: Personalized federated learning with meta-learning. In: Advances in Neural Information Processing Systems (NeurIPS) (2020)
6. Fuchs, M., Fischer, L., Montuoro, A., Kumar, M., Moser, B.A.: Risk assessment in ai system engineering: Experiences and lessons learned from a practitioner's perspective. In: Database and Expert Systems Applications - DEXA 2024 Workshops, pp. 67–76. Springer Nature Switzerland, Cham (2024). https://doi.org/10.1007/978-3-031-68302-2_6
7. Gao, C., Wang, Q., Wu, X., Zhang, Z.J.: Winclip: Efficient fine-tuning of clip for fine-grained visual recognition. arXiv preprint arXiv:2306.16811 (2023)
8. Gou, J., Yu, B., Maybank, S.J., Tao, D.: Knowledge distillation: a survey. Int. J. Comput. Vis. **129**(6), 1789–1819 (2021)
9. Grativol, L., Leonardon, M., Muller, G., Fresse, V., Arzel, M.: Flocora: federated learning compression with low-rank adaptation. In: 2024 32nd European Signal Processing Conference (EUSIPCO), pp. 1786–1790. IEEE (2024)
10. Hu, E.J., et al.: Lora: Low-rank adaptation of large language models. ICLR **1**(2), 3 (2022)
11. Jung, C., Kim, M., Hwang, W., et al.: Anomalyclip: Unsupervised anomaly detection and localization via vision-language prompting. arXiv preprint arXiv:2306.16926 (2023)
12. Kumar, M., et al.: Geometrically inspired kernel machines for collaborative learning beyond gradient descent (2024), https://arxiv.org/abs/2407.04335
13. Kuo, K., Raje, A., Rajesh, K., Smith, V.: Federated lora with sparse communication. arXiv preprint arXiv:2406.05233 (2024)
14. Li, T., Sahu, A.K., Talwalkar, A.: Ditto: fair and robust federated learning through personalization. In: Proceedings of the 40th International Conference on Machine Learning (ICML) (2023)
15. Liu, H., Li, C., Wu, Q., Lee, Y.J.: Visual instruction tuning. Adv. Neural. Inf. Process. Syst. **36**, 34892–34916 (2023)
16. Mangrulkar, S., Gugger, S., Debut, L., Belkada, Y., Paul, S., Bossan, B.: Peft: State-of-the-art parameter-efficient fine-tuning methods (2022). https://github.com/huggingface/peft
17. McMahan, B., Moore, E., Ramage, D., Hampson, S., y Arcas, B.A.: Communication-efficient learning of deep networks from decentralized data. In: Artificial intelligence and statistics. pp. 1273–1282. PMLR (2017)
18. Peng, Z., et al.: Kosmos-2: grounding multimodal large language models to the world. arXiv preprint arXiv:2306.14824 (2023)
19. Radford, A., et al.: Learning transferable visual models from natural language supervision. In: International Conference on Machine Learning, pp. 8748–8763. PmLR (2021)
20. Redmon, J., et al.: You only look once: Unified, real-time object detection. In: IEEE Conference on Computer Vision and Pattern Recognition (CVPR) (2016)
21. Ren, S., He, K., Girshick, R., Sun, J.: Faster r-cnn: towards real-time object detection with region proposal networks. Adv. Neural Inform. Process. Syst. (NeurIPS) **28** (2015)
22. Ro, S., Satwika, S., Gayathri, P.Y., Balsha, M.G., Ucar, A.: Fine-tuning florence2 for enhanced object detection in un-constructed environments: vision-language model approach. arXiv preprint arXiv:2503.04918 (2025)
23. Sattler, F., Mueller, K.R., et al.: Clustered federated learning: model-agnostic distributed multitask optimization under privacy constraints. IEEE Trans. Neural Netw. Learn. Syst. (2020)

24. Sun, Y., Li, Z., Li, Y., Ding, B.: Improving lora in privacy-preserving federated learning. arXiv preprint arXiv:2403.12313 (2024)
25. Sun, Y., et al.: Partialfed: Cross-subset personalized federated learning. arXiv preprint arXiv:2110.00620 (2021)
26. Tan, A.Z., Yu, H., Cui, L., Yang, Q.: Towards personalized federated learning. IEEE Trans. Neural Netw. Learn. Syst. **34**(12), 9587–9603 (2022)
27. Vahidian, A., et al.: Efficient clustered federated learning via signature-guided client grouping. In: Proceedings of the 37th AAAI Conference on Artificial Intelligence (2023)
28. Xiao, B., et al.: Florence-2: advancing a unified representation for a variety of vision tasks. In: Proceedings of the IEEE/CVF Conference on Computer Vision and Pattern Recognition, pp. 4818–4829 (2024)
29. Ye, D., et al.: Federated learning: a comprehensive survey of challenges and solutions. ACM Comput. Surv. (2023)
30. Yi, L., Yu, H., Wang, G., Liu, X., Li, X.: pfedlora: model-heterogeneous personalized federated learning with lora tuning. arXiv preprint arXiv:2310.13283 (2023)
31. Zhang, C., et al.: Federated adaptation for foundation model-based recommendations. arXiv preprint arXiv:2405.04840 (2024)
32. Zhuang, W., Chen, C., Lyu, L.: When foundation model meets federated learning: Motivations, challenges, and future directions. arXiv preprint arXiv:2306.15546 (2023)

Efficient Federated Learning Integration into Existing MLOps Pipelines via Centralized Model Management

Tatjana Krau[✉], Florian Huber, Teena Chirakal, Tobias Ricken, Bernd Lüdemann-Ravit, and Frieder Heieck

Institue of Production and Informatics, Sonthofen, Germany
{tatjana.krau,florian.huber,teena.chirakal,tobias.ricken,
bernd.luedemann-ravit,frieder.heieck}@hs-kempten.de

Abstract. Federated Learning (FL) offers a solution to the challenges of traditional centralized machine learning by enabling decentralized training and exchanging only model updates instead of raw data. This approach addresses key issues such as privacy concerns and high data transfer costs. However, integrating FL into existing Machine Learning Operations (MLOps) pipelines presents challenges, particularly regarding model versioning, synchronization, and scalability. This paper introduces a concept for centralized model management that enables the integration of FL into existing MLOps pipelines without the need to overhaul the existing architecture. The concept is specifically developed for deployment in an industrial setting, with plans for implementing both FL and Transfer Learning (TL) in the future. The proposed approach emphasizes flexibility, ensuring that it can be easily extended to accommodate additional methods and seamlessly integrated into diverse, pre-existing infrastructure. The management of the system is facilitated using the open-source tool MLflow, which offers significant advantages over specialized FL frameworks, particularly in terms of adaptability and resource optimization.

Keywords: MLOps · Federated Learning · Model Management

1 Introduction

Traditional Machine Learning (ML) approaches face significant limitations like centralized training, which is the training of a single model based on raw input data. This often leads to a lack of generalizability, as they are tailored to the specific characteristics of single data sources and fail to adapt to diverse use cases or the varying contexts across different companies. This challenge, along with regulatory limits on data sharing, highlights the need for new methods that can effectively use decentralized data to overcome the weaknesses of traditional approaches. FL has emerged as a transformative paradigm to address these limitations, enabling collaborative model training across decentralized devices without the need to exchange sensitive data [4]. In this methodology, each so-called client train a *local model* on its own dataset and transmits only model updates, such as weight adjustments, to a central server, seen in Fig. 1. These updates are aggregated to create a global model that benefits from diverse data sources while maintaining privacy [2].

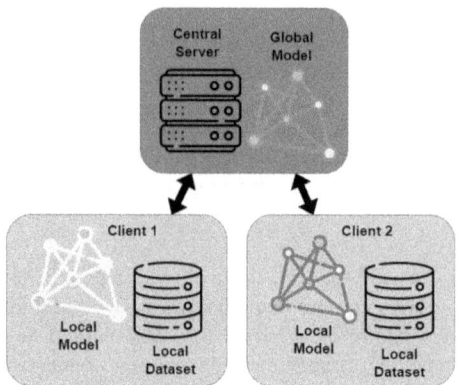

Fig. 1. FL workflow: Local models are aggregated on a central server to a global model, which is then sent back to each client.

By keeping data localized and sharing only model updates, FL offers significant advantages over classical ML, including enhanced privacy preservation, reduced data transfer costs, and the ability to leverage diverse, real-world data sources.

These benefits have fueled its growing importance in domains such as healthcare, finance, and edge computing, where data sensitivity and distribution are key aspects [1,6].

Despite that, FL introduces distinct challenges which complicate practical deployment. Heterogeneous data across participating nodes—stemming from differences in device capabilities, data distributions, or collection methods—can degrade model performance and convergence. Data synchronization poses another hurdle, as asynchronous updates from distributed clients must be effectively coordinated to maintain consistency across models. Furthermore scalability remains a critical concern, as FL systems might accommodate thousands of participants while managing computational and communication overheads [1,2]. Perhaps most crucially, model management in FL is inherently challenging, requiring robust mechanisms to track, version, and integrate updates from various sources into a cohesive global model. These challenges underscore the need for structured approaches to operationalize FL effectively.

In this context, MLOps can provides a foundational framework to streamline the integration of FL into practical workflows. MLOps encompasses a set of practices and tools designed to automate and manage the end-to-end lifecycle of ML models, from development and training to deployment and monitoring. Its importance lies in its ability to enhance reproducibility, scalability, and collaboration in ML systems, making it a critical for enabling enterprise-grade solutions [5,10]. Extending MLOps pipelines through, for instance, centralized model versioning and management instances, allows FL to leverage existing infrastructure to address its unique demands. Such pipelines offer a systematic approach to orchestrate distributed training, synchronize model updates, and maintain a unified model registry, thereby bridging the gap between FL's decentralized nature and the operational rigor required for real-world applications.

Given the transformative potential of FL and its inherent challenges, a critical question emerges: How can FL be effectively integrated into existing MLOps pipelines? Leveraging established architectures offers several compelling advantages. First, it reduces implementation costs by utilizing existing resources, such as computational infrastructure and trained personnel, rather than requiring entirely new systems. Second, it facilitates seamless integration into organizational structures, aligning FL workflows with established processes for model development, deployment, and governance. Third, it ensures consistency across MLOps, maintaining standardized practices that enhance reliability and interoperability. By building on proven MLOps frameworks, organizations can accelerate FL adoption while minimizing disruption and resource overhead.

This paper focuses on the efficient integration of FL into existing MLOps pipelines by extending a central model versioning and management platform, rather than building a new architecture from scratch. Specifically, we use MLflow[1], an open-source platform that optimizes the machine learning lifecycle by managing experiments, tracking model versions, and facilitating deployment.

The concept is grounded in a practical use case, aiming to implement FL in an industrial machine park without the need for changes to the existing IT infrastructure. By utilizing vibration data from the machines, this approach enables real-time wear detection and optimized maintenance while ensuring compatibility with the current system setup. After implementing FL, TL will be applied within the same machine park. TL allows pre-trained models to be efficiently adapted to new machines or tasks, significantly reducing training effort [13]. In order to support the flexible use of both methods, a comprehensive model management concept has been created to integrate FL and TL seamlessly.

To validate the practicality of this approach, two well-established MLOps architectures have been implemented as a lab setup at our institute. This setup serves to validate the practical applicability and functionality of the concept before deploying it in an industrial environment.

Although this concept is demonstrated in the context of wear detection using vibration data, the underlying model management framework is flexible and can be adapted to various use cases. Whether the model is trained on vibration data for wear detection or other types of sensor data for anomaly detection, the structure of the management system remains applicable and scalable across different domains.

---

[1] https://mlflow.org/.

In light of these points, the following research questions guide the investigation of how this concept can be effectively applied and optimized:

- **RQ1:** How can FL be efficiently incorporated into existing MLOps pipelines to optimize performance and resource use without disrupting current workflows?
- **RQ2:** Which specific benefits does MLflow provide over specialized FL frameworks in terms of flexibility, scalability, and operational efficiency, particularly when integrated with TL to enable adaptation across diverse machine types and tasks in industrial environments?

The remainder of this paper is structured as follows. Sect. 2 explores frameworks for model versioning and integration of TL methods, with a particular focus on the use of MLflow. Sect. 3 presents the proposed architecture of the complete MLOps pipeline integrated with FL, while Sect. 4 elaborates on the concept of model management within this architecture. Finally, Sect. 5 provides a conclusion and an outlook on future developments and potential applications of the proposed system.

2 Related Work

FL has established itself as a commonly-used method for decentralized ML. The practical implementation of FL can be achieved through either specialized FL frameworks that provide mechanisms for distributed training and aggregation or by custom-build solutions. Frameworks such as Flower[2], IBM FL[3] and WrapperFL provide a structured environment for FL experiments [9,11]. However, these are often strongly geared towards specific scenarios and require more extensive customization of the IT infrastructure. Comparisons show that open-source solutions such as Flower allow for greater flexibility, while proprietary solutions often offer better security and data protection mechanisms [9].

The combination of FL with MLOps is increasingly seen as essential for the efficient management and scaling of FL projects. Due to its distributed nature and the need for efficient model management, FL places special demands on infrastructure. While classical MLOps approaches focus on centralized ML models, FL requires flexible management of multiple local models, global models and dynamic scaling of model distribution [8].

Moon et al. [8] show that a MLOps-like methodology can be specifically applied to FL by covering the entire lifecycle of a FL model. A crucial point here is the management of client heterogeneity, as FL systems have to work with differently clients and varying data distributions.

Malyuk [7] shows that existing MLOps concepts can be extended for FL by using DevOps, containerization and orchestration techniques. This allows FL to be deployed automatically and managed efficiently without the need to completely redevelop specialized FL frameworks.

---

[2] https://github.com/adap/flower.
[3] https://github.com/IBM/federated-learning-lib.

The integration of model management tools is already a central component of classical MLOps and has proven to be a key factor in increasing the reproducibility, the scalability and traceability of ML experiments [12].

Common model management tools include Weight&Biases[4], DVC[5], Neptune.ai[6] and MLflow, which provide mechanisms for experiment tracking, model versioning and scaling, but they differ in their focus, functionality and ease of use [3]. A comparative analysis shows that MLflow has established itself as particularly suitable for use in MLOps due to its open source availability, modular architecture and simple integration with existing IT infrastructures. Although MLflow was primarily developed for centralized ML models, it offers a flexible and extensible solution for the management of distributed FL models [12]. The decision in favour of MLflow as a model management tool for FL is based on several factors:

- **Widespread use:** MLflow is an established tool with a large community and a steadily growing user base.
- **Flexibility:** The tool can be seamlessly integrated into existing MLOps pipelines, supports multiple programming languages and can be operated both locally and in the cloud environement.
- **Cost savings:** Since MLflow is open source, there are no additional license fees, as would be the case with proprietary solutions.
- **Extensibility:** The modular architecture allows the integration of FL-specific methods without the need to switch to specialized FL frameworks. Due to this feature, MLflow can also be used for other methods such as TL.

3 Architecture

Modern industrial processes increasingly rely on ML for data-driven optimization, with models being deployed directly on edge devices to enable on-site processing. A typical MLOps pipeline consists of multiple stages, including data collection and storage, model training, and deployment. This section provides an overview of a standard MLOps infrastructure and explores how it can be extended for FL. Figure 2 illustrates the typical workflow, with the left side showing well-known MLOps pipeline variants in industry that have been implemented at our institute. These serve as the foundation for the future integration of FL. The right side presents the proposed concept, which will be gradually incorporated into the existing infrastructure.

The first step is data collection, for example through sensors. **Data availability** is crucial, as reliable and continuous data collection forms the foundation for all subsequent steps. An **edge device** retrieves this data, optionally

---

[4] https://wandb.ai/site.
[5] https://dvc.org/.
[6] https://neptune.ai/.

pre-processes it, and transmits it to a central database for storage. In our implementation, we use a Welotec EG50 Mk2[7] edge device, which runs a Python script for data acquisition and transmission.

After data collection, we distinguish between two commonly used approaches, selected based on their widespread adoption in the industry:

- **On-Premise Solution (Yellow):** Data is stored on a local server. Model training takes place on a Virtual Machine (VM) and includes methods such as wear or anomaly detection. MLflow is used for managing the model, training parameters, evaluation metrics, and metadata.
- **Cloud-Based Solution (Red):** Data is stored in the Azure Cloud, model training is conducted using Databricks, and model management is handled via MLflow, which also runs in the cloud.

Regardless of the chosen pipeline, the final model is typically deployed directly onto the edge device in traditional ML workflows. However, when utilizing FL, the architecture is extended as follows, seen in the left side of Fig. 2:

- Locally trained models from each process are stored in a **central model management instance**.
- A **FL method** computes a global model with the given local models.
- The global model, like the local models, is stored in the central model management system along with relevant training parameters and evaluation metrics. Detailed information about the structure of the model management follows in Sect. 4.
- The global model is then deployed across all edge devices for production use.

This cycle repeats continuously, improving the global model with newly trained or added local models. Both the central model management system and the FL method can be implemented either on-premise or in the cloud, as indicated by the yellow and red markings in Fig. 2. It is important to note that the existing MLOps pipeline for conventional ML remains unchanged. The only additions are the central model management system and the FL method (right half of Fig. 2). However, a well-structured model management system is essential, which, in our implementation, is managed using MLflow. While the choice of a specific FL method is an important consideration, it is beyond the scope of this paper. The management of MLflow is discussed in Sect. 4.

4 Conceptual Design

Based on the context that this concept is developed for the future implementation of FL on machining tools for wear detection, we consider two machines exemplary, which will initially train local models. These local models will later be aggregated to create a global model through FL.

---

[7] https://welotec.com/de/products/eg500-mk2.

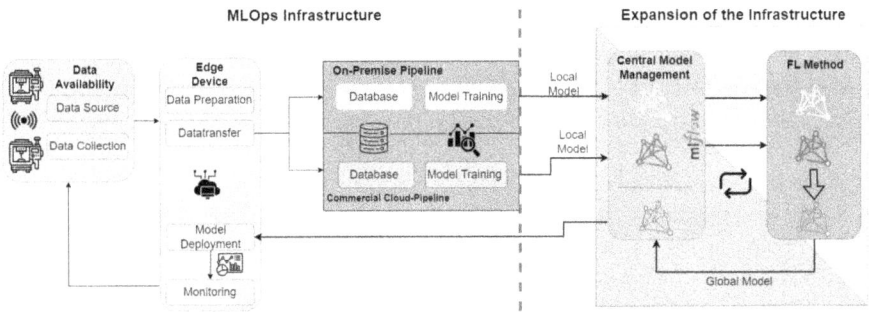

Fig. 2. MLOps pipeline consisting of the components data availability, edge device for data transfer, data storage and model training with an extension for integrating FL. Distinction between an on-premise solution (yellow) and a cloud-based solution (red). The extension consists of central model management and the implementation of the FL method. (Color figure online)

The model management system consists of three key components: the storage of the local models from the two machines along with the associated training data, the global model and its metadata, and the management of the TL methodology. Each of these components is detailed in the following sections and is outlined in Table 1.

This structure ensures that the integration of FL and TL into existing systems is streamlined and efficient, maintaining flexibility for future extensions and adaptations.

1. **Local Models of Machines A and B with Metadata**
   Each machine trains its own model using a specific dataset. In MLflow, the models are associated with metadata that includes essential information about the training parameters and conditions. This includes the learning rate, batch size, timestamp of the data used, and performance metrics such as accuracy. The detailed documentation of these parameters is crucial for model traceability, evaluation, model comparability, and reproducibility. The choice of metadata to be stored depends on the method and application, and it is crucial to carefully determine which information should be recorded. But they can be expanded if further information is needed in a specific scenario. The model is stored in the artifacts. It is very important to give an unique name to each model to avoid confusion. It is marked with color in the table.

2. **Usage of Models in the FL Method**
   Once the local models are trained, they are used in the FL method. The models from Machine A and Machine B are loaded and used to train a global model. It is important to document the local models used in the metadata to track which models contributed to the global aggregation. This information is stored in the tags marked with the specific color of model A and B, making the entire training process traceable. Additionally, the parameters of the FL

method, such as the aggregation method used, are recorded. The final global model is then saved in artifacts and is made accessible to both machines.

3. **Extension through TL**
The architecture is further extended by the introduction of TL. A new training is set up as a TL experiment in MLflow. The models, data, and transfer method used are documented, along with the performance metrics as before for FL experiments. It is also important to tag the used models highlighted with color in the table.

The presented concept for management in MLflow enables flexible expansion of existing architectures and provides an easy way to integrate not only FL but also TL.

Table 1. Illustration of the proposed model management structure within MLflow, highlighting key components for versioning for FL and TL.

Experiment	Machine A	Machine B
Run	Local_Training_A_1	Local_Training_B_1
Parameter	learning rate, batch size	learning rate, batch size
Metrics	accuracy, loss	accuracy, loss
Artifacts	client_A_model_v1.pth	**client_B_model_v1.pth**
Tags	Device_ID, timestamp_data	Device_ID, timestamp_data
Experiment	Federated Learning	
Run	Global_Aggregation_A_B_1	
Parameter	aggregation_method, num_clients, name_clients	
Metrics	accuracy, loss	
Artifacts	global_model_v1.pth	
Tags	used_models=client_A_model_v1.pth, **client_B_model_v1.pth**	
Experiment	Transfer Learning	
Run	TL_Adaption_A_1	
Parameter	transfer_method	
Metrics	accuracy, loss	
Artifacts	transfer_model_v1.pth	
Tags	used_models =**client_B_model_v1.pth**, used_data, Device_ID	

5 Conclusion and Future Work

This paper presents a concept that enables the seamless integration of FL into existing MLOps pipelines—whether on-premise or cloud-based—by building a model management structure within MLflow. The integration of FL into existing pipelines can be achieved efficiently through MLflow's flexible model management framework. This approach ensures minimal disruption to current workflows, while optimizing resource use, particularly in decentralized environments. Furthermore, MLflow offers significant advantages over specialized FL frameworks, particularly in terms of flexibility, scalability, and operational efficiency. The integration of FL with TL within this structure enhances the ability to adapt across diverse machine types and tasks, which is essential in industrial settings.

Future work will focus on evaluating the approach in a real-world use case using the developed MLOps pipeline with the edge device from Welotec. Once the approach has been proven effective, it will be expanded to the industrial applications mentioned in Sect. 1.

Further challenges include examining scalability with respect to the number of clients and potential infrastructure limitations. Additionally, it must be investigated whether fine-tuning the experiment tracking is necessary, particularly regarding the storage of additional data, which depends on the specific use case. Another key aspect is the implementation and comparison of different FL methods and frameworks to analyze the advantages and disadvantages of custom implementations versus existing solutions. A critical factor that must also be addressed is the synchronization of models, which is essential for ensuring consistent and accurate global updates. For real-world deployment, it would be beneficial to implement a client manager to handle synchronization, thus ensuring smooth operation within the federated architecture. Finally, security considerations must be addressed, particularly regarding data privacy and secure communication within the federated architecture.

Acknowledgment. This research was supported by the *Bavarian Ministry of Economic Affairs, Regional Development and Energy* under the project FLInK[1] (FI: DIK-2308-0031// DIK0537/03). We are grateful for their support and funding, which made this work possible. Additionally, we would like to thank Welotec GmbH for providing the edge device, which was essential for our experiments([1]https://kefis.fza.hs-kempten.de/de/forschungsprojekt/585-flink

References

1. Banabilah, S., Aloqaily, M., Alsayed, E., Malik, N., Jararweh, Y.: Federated learning review: Fundamentals, enabling technologies, and future applications. Inform. Process. Manag. **59**(6), 103061 (2022). https://doi.org/10.1016/j.ipm.2022.103061, https://www.sciencedirect.com/science/article/pii/S0306457322001649
2. Bhanbhro, J., Nisticò, S., Palopoli, L.: Issues in federated learning: some experiments and preliminary results. Sci. Rep. **14**(1), 29881 (Dec 2024). https://doi.org/10.1038/s41598-024-81732-0, https://www.nature.com/articles/s41598-024-81732-0,
3. Idowu, S., Osman, O., Strüber, D., Berger, T.: Machine learning experiment management tools: a mixed-methods empirical study. Empirical Softw. Eng. **29**(4) (2024). https://doi.org/10.1007/s10664-024-10444-w

4. Kairouz, P., McMahan, H.B., Avent, B., Bellet, A., Bennis, M., Bhagoji, A.N.e.a.: Advances and Open Problems in Federated Learning (Mar 2021). https://doi.org/10.48550/arXiv.1912.04977
5. Kreuzberger, D., Kühl, N., Hirschl, S.: Machine Learning Operations (MLOps): Overview, Definition, and Architecture. IEEE Access **11**, 31866–31879 (2023). https://doi.org/10.1109/ACCESS.2023.3262138, https://ieeexplore.ieee.org/document/10081336/?arnumber=10081336, conference Name: IEEE Access
6. Li, L., Fan, Y., Tse, M., Lin, K.Y.: A review of applications in federated learning. Comput. Indust. Eng. **149**, 106854 (Nov 2020). https://doi.org/10.1016/j.cie.2020.106854, https://www.sciencedirect.com/science/article/pii/S0360835220305532
7. Malyuk, A.: FLOps: Practical Federated Learning via Automated Orchestration (on the Edge)
8. Moon, J., Yang, S., Lee, K.: FedOps: a platform of federated learning operations with heterogeneity management. IEEE Access **12**, 4301–4314 (2024). https://doi.org/10.1109/ACCESS.2024.3349691, https://ieeexplore.ieee.org/document/10380563/?arnumber=10380563
9. Riedel, P., Schick, L., von Schwerin, R., Reichert, M., Schaudt, D., afner, A.: Comparative analysis of open-source federated learning frameworks - a literature-based survey and review. Inter. J. Mach. Learn. Cyberne, **15**(11), 5257–5278 (2024). https://doi.org/10.1007/s13042-024-02234-z
10. Testi, M., et al.: MLOps: A Taxonomy and a Methodology. IEEE Access **10**, 63606–63618 (2022). https://doi.org/10.1109/ACCESS.2022.3181730, https://ieeexplore.ieee.org/document/9792270/?arnumber=9792270 conference Name: IEEE Access
11. Wu, X., Tan, S., Xu, Q., Yang, Q.: WrapperFL: A Model Agnostic Plug-in for Industrial Federated Learning (Aug 2022). https://doi.org/10.48550/arXiv.2206.10407
12. Zaharia, M., et al.: Accelerating the Machine Learning Lifecycle with MLflow. IEEE Data Eng. Bull. (2018). https://www.semanticscholar.org/paper/Accelerating-the-Machine-Learning-Lifecycle-with-Zaharia-Chen/b2e0b79e6f180af2e0e559f2b1faba66b2bd578a
13. Zhuang, F., et al.: A Comprehensive Survey on Transfer Learning (Jun 2020). https://doi.org/10.48550/arXiv.1911.02685

Deep Photometric Stereo for Tool Wear Inspection

Thomas Jäkel(✉) and Frank Schirmeier

Institute for Data-Optimized Manufacturing (IDF), University of Applied Sciences Kempten,
Bahnhofstraße 61, 87435 Kempten, Germany
`thomas.jaekel@hs-kempten.de`

Abstract. This paper explores the potential of a modern AI-based photometric stereo model for detecting and assessing tool wear in machining. A retrofitted microscope is used to scan a test workpiece with a well-defined geometry, allowing for a quantitative evaluation of the resulting normal map. Additionally, a qualitative analysis is conducted on a scan of an end mill. Beyond normal maps, the study discusses the integration of additional Bidirectional Reflectance Distribution Function (BRDF) maps for the assessment of tool wear.

Keywords: Photometric Stereo · Tool Wear · Machining

1 Introduction

The detection and assessment of tool wear represent a central challenge in modern, automated production environments for subtractive manufacturing [1]. To prevent rejects caused by dimensional inaccuracies or poor surface finish, tools are usually assigned predefined service live times or cutting distances, with a large safety factor incorporated due to the high variability in a tool's lifespan. As a result, tools are often replaced prematurely before reaching their actual lifespan, leading to avoidable costs and environmental impact. Since the manual assessment of milling or turning tools is time-consuming, error-prone, and highly subjective, various approaches exist for the automated evaluation of tool condition. These can generally be divided into direct and indirect methods.

Indirect methods are characterized by evaluating various process parameters to assess the tool's condition instead of directly examining the tool itself. These process parameters can be obtained from the machine control system or from externally mounted sensors [2]. Examples include the measurement and evaluation of vibrations [3], acoustic emissions [4], spindle motor currents [5], or cutting forces [6]. The advantages of indirect measurement include the fact that no intervention in the process is required, leading to time savings and the possibility of real-time tool diagnostics and tool life prediction. However, challenges in indirect measurement mainly arise in data collection and analysis. Often, even when using costly additional sensors, small wear phenomena are overshadowed by other effects in the measurement chain or lost in noise. Moreover, establishing tool-independent criteria for assessing wear is a central challenge in interpreting measurement data. Modern approaches use AI algorithms [7] to analyse high-dimensional data from various sources [8] to diagnose the tool's condition.

Direct methods, on the other hand, are characterized by evaluating the tool itself. These methods generate two- or three-dimensional data of the tool in a usually contactless manner [9] to determine its actual condition. In two-dimensional analysis, image data is typically generated and then evaluated using classical image processing [10] or AI algorithms [11]. The challenge in capturing these images is to provide wear phenomena in a quality sufficient for the chosen algorithm. Reflections caused by lighting or by the environment—especially on polished tools or due to surface coatings—make data collection more difficult. Studies in this field explore methods for capturing images without reflections [12] and improving algorithms to make image analysis more robust against reflections or contaminants. A further research focus is the three-dimensional assessment of tool wear, driven in particular by the development and commercialization of technologies such as focus variation [13, 14], confocal microscopy [15], white light interferometry [16], and profile lasers [17]. Additionally, simpler techniques, such as contour mapping with an analog camera and a shallow depth of field, have been used since 1966 to generate three-dimensional information on tool wear [18].

Another method for capturing three-dimensional (2.5D) data is photometric stereo [19]. In this technique, the object is illuminated from different angles to reconstruct surface normals. While the resulting normal map itself does not contain full 3D geometry, true three-dimensional data can be obtained through normal integration or depth estimation. This method is already used in industry, for example, in automated defect detection and assessment [20–22], or surface roughness inspection [23]. However, classical mathematical approaches are typically limited to Lambertian surfaces, making them unsuitable for analyzing cutting tools. The development of novel AI algorithms, which are robust against specular reflecting surfaces, opens up new application possibilities. This paper evaluates the application of a Deep photometric stereo network for assessing tool wear.

1.1 Tool Wear

Tool wear is a critical factor in subtractive machining, as it directly impacts key parameters such as geometric accuracy, surface finish or process stability. Cutting tools in machining are simultaneously subjected to various mechanical, thermal, and chemical stresses. These stresses lead to different wear mechanisms [24] and, ultimately, different types of wear. A description of the wear deterioration forms of end mills and their measurement can be found in ISO 8688-2 [25], which focuses on tool life testing. These forms are illustrated in Fig. 1 (a–m), with built-up edge shown separately as (n), as it is typically not classified as a tool wear process but can contribute to other wear forms. The type of wear used as a tool change criterion can vary depending on the specific process, tool, and material combination or the focus of the analysis [26]. One criterion that is often used is flank wear, by which different assessment strategies and quantitative measures can also be applied [27].

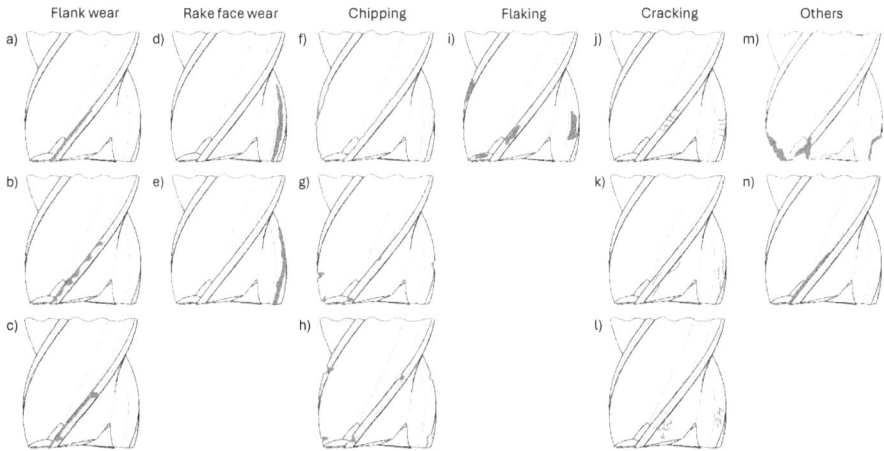

Fig. 1. Tool deterioration: a) uniform flank wear; b) non-uniform flank wear; c) localized flank wear; d) crater wear; e) stair-formed face wear; f) uniform chipping; g) non-uniform chipping; h) localized chipping; i) flaking; j) comb cracks; k) parallel cracks; l) irregular cracks; m) catastrophic failure; n) built up edge (own depiction based on the tool wear types described in ISO 8688-2 [25])

1.2 Photometric Stereo

The fundamental idea behind photometric stereo is to estimate an object's surface normal (vectors that are perpendicular to a surface) by analyzing its appearance under multiple lighting directions while keeping the viewpoint fixed. The position of the light source is either assumed to be known (calibrated approaches) or unknown (uncalibrated approaches). The result is often visualized as a normal map, where colour represents the normal vector. The red channel (0 to 255) corresponds to the X-axis, the green channel (0 to 255) represents the Y-axis, and the blue channel (128 to 255) encodes the Z-axis, which always points towards the viewer. The classical mathematical approach is mostly limited to Lambertian surfaces, where the relationship between the incoming light ($\vec{l}$), surface normal ($\vec{n}$), and outgoing light ($\vec{v}$) as seen in Fig. 2 can be easily described and solved. [19].

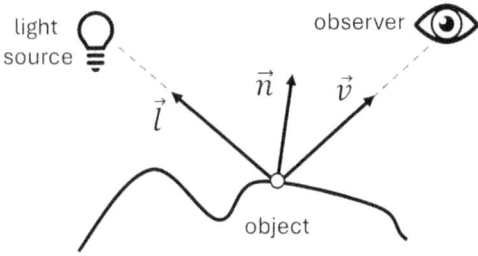

Fig. 2. Photometric Stereo

In reality, however, additional effects such as specular reflections, interreflections, light refraction, or subsurface scattering come into play. In the case of cutting tools, specular reflections are particularly relevant due to the choice of material, surface finish, or tool coating. For such surfaces, the simple mathematical approach for Lambertian surfaces is no longer valid, requiring either significantly more complex mathematical models or AI-based algorithms.

For this study, we chose SDM-UniPS [28]. Unlike other models it does not require or assume specific surface properties, geometries, or lighting models, removes the need for masks in many scenarios, handles high-resolution images, and improves detail preservation even under challenging lighting conditions. By combining scale-invariant encoding, non-local feature interaction, and diverse training data, it outperforms both calibrated and uncalibrated methods, making it practical for real-world, non-commercial applications. Additionally, the model provides not only surface normals but also BRDF parameters (base colour, roughness, and metallic properties), which can be used as an additional data source for detecting and assessing wear phenomena.

2 Results

The assessment of Photometric Stereo models under general conditions has already been addressed by other publications [29]. The aim of this study is to determine whether the chosen model, with this experimental setup, is suitable for meeting the technical requirements in the assessment of tool wear. To achieve this, two approaches are used: one involves a test workpiece with a known geometry, and the other uses a worn an end mill with unknown geometry. The first approach was chosen because the exact geometry of an end mill is highly complex and difficult to replicate, making it challenging to create a "ground truth." To still evaluate the results on the complex geometry of an end mill, the second approach was chosen, focusing on a qualitative assessment of the model. In future work, collaboration with a tool manufacturer could provide the exact tool geometry, enabling the development of high-quality "ground truth" datasets. A logical and highly relevant next step would be the development of automated tool wear regression models, though this falls beyond the scope of the present study.

2.1 Materials and Methods

For the capture of images, an older, modified measuring microscope 06307 MM from Mahr OKM GmbH (now Mahr GmbH) was used. Microscopes of this type are sold by various manufacturers and are widely used in industry. The measuring microscope retained its original optics, which include the Navitar 1-6265D lens, the 1-6218 1X adapter, and the 1-6010 C-Mount coupler. The original camera of the microscope has been replaced with the "High Quality Camera" from the Raspberry Pi Foundation, which can be mounted without requiring further modifications. This camera is equipped with a 12.3-megapixel sensor, the IMX477 from Sony, with a diagonal of 7.857 mm (1/2,3") and square pixels. This solution offers the advantage of camera control—such as adjusting exposure time—via the Raspberry Pi (model 4 B in our case) using Python code. This enables seamless automation of the image acquisition and data collection process (Fig. 3).

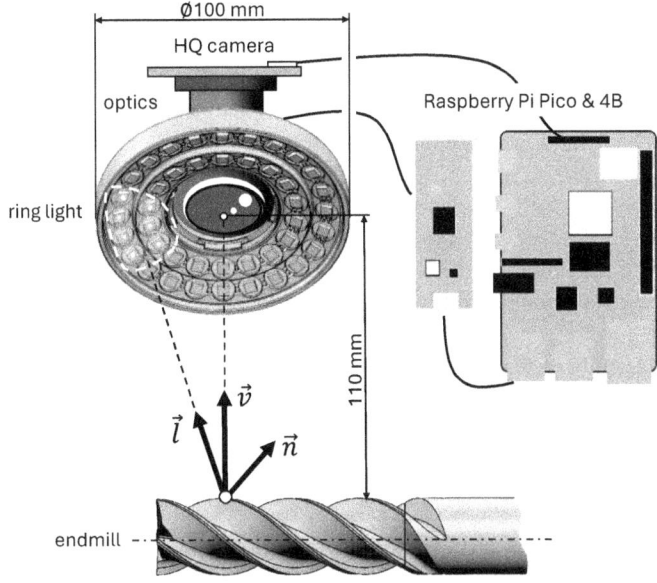

Fig. 3. Measurement Setup

A custom-designed, ring light using SK6812 LEDs, controlled by a Raspberry Pi Pico, was employed for illumination. The LEDs are individually addressable, offering the ability to turn them on or off one by one, as well as to adjust their brightness and colour (RGBW). This setup allows the creation of light zones that can illuminate the endmill from various angles. The ring light has an outer diameter of approximately 100 mm and an inner diameter of 41 mm, with a working distance of 110 mm. The inner LEDs are angled (15.5°) toward the focal point, while the outer LEDs are parallel to them, directing light slightly outward from the focal point. Similar lighting strategies are used in modern measuring microscopes to enhance the visibility of surface features such as defects, including scratches or milling marks. Ring lights with zonal illumination are also commercially available, providing a straightforward and cost-effective implementation of the presented concept.

2.2 Evaluation Workpiece

The test workpiece is a milled cylindrical component with a diameter of 6 mm, made from aluminum 7075. It features a circumferential $1 \times 45°$ chamfer, a 0.5 mm step with an adjacent 45° bevel, and a 1 mm-wide groove with vertical sidewalls. The surface remained untreated, meaning it was neither coated nor polished.

Aluminum exhibits high reflectivity of approximately 90%, depending on the wavelength. This strong reflectivity lead to intense internal reflections within the groove, resulting in in the highest angular variations at this location. Similar reflections can also occur in the chip space between the flutes of end mills, making it crucial to give extra attention to these areas.

Figure 4 displays the ideal normal map of the evaluation workpiece, generated from the CAD model, alongside the normal maps produced by the SDM-UniPS model and the classical Woodham model, which assumes a Lambertian surface with a known light source. Additionally, a heatmap illustrates the angular differences between the model outputs and the ideal normal map for both models. The input data for both models include nine images of lightning from different directions, along with a mask used to evaluate only the workpiece.

The SDM-UniPS model achieves a mean angular difference of 17.15°, with a maximum angular error of 89.77°. In comparison, the Woodham model performs less accurately, with a mean angular difference of 54.15° and a maximum angular error of 107.57°.

The high angular errors at the edges of the workpiece may also result from slight misalignments between the ideal normal map and the model's normal map or dimensional variation between the ideal CAD model and the milled part, which is inevitable.

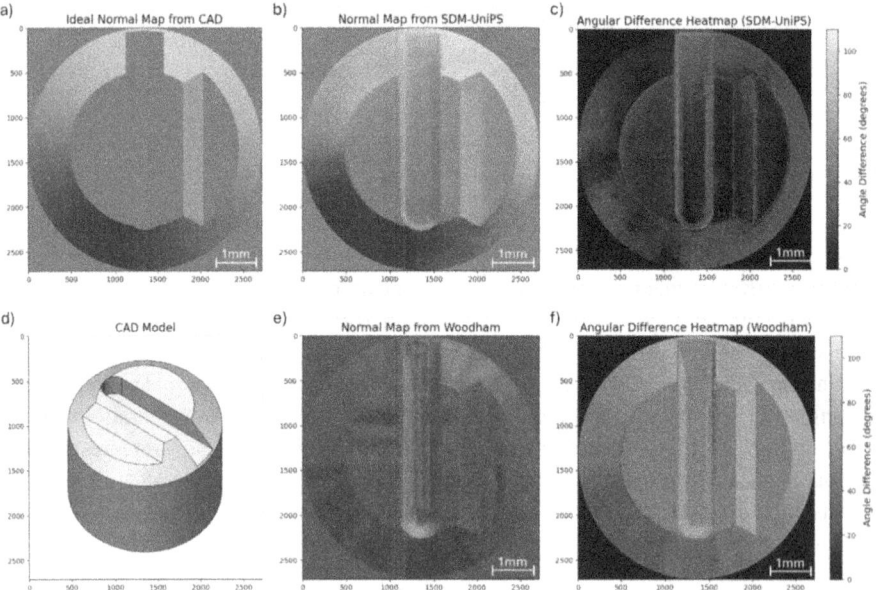

Fig. 4. Evaluation workpiece: a) Ideal Normal Map from CAD; b) Normal Map from SDM-UniPS; c) Angular Difference Heatmap (SDM-UniPS); d) CAD Model; e) Normal Map from Woodham; f) Angular Difference Heatmap (Woodham)

2.3 Endmill Scan

The scanned endmill is a CEPR6060-TH model from MOLDINO Tool Engineering, Ltd. It is made of carbide, has a 6 mm diameter, six flutes, and features a TH layer—a PVD (Physical Vapor Deposition) coating with a distinctive copper-brown colour. The scanned example exhibits significant wear, including extensive flank wear, chipping, and flaking.

As with the test component, the input data consists of nine photos (Fig. 5(a)) taken under different lighting directions using the ring light zones. The images were pre-cropped to a square format (3040 × 3040 Pixels), which was found to enhance overall quality. Given this resolution, a single pixel corresponds to approximately 0.002 mm. Defects as small as 0.01 mm were clearly visible in the resulting normal map, making it suitable for monitoring minor wear phenomena during the milling process, as demonstrated in [30]. No mask was used as a model input; while masking can improve quality in certain areas, we found that it may also introduce deterioration in others.

A common challenge in microscope photography is the shallow depth of field, which can create issues for SDM-UniPS due to blurred regions. Additionally, the strong cast shadows in the images could potentially affect the results. To optimize GPU memory usage, the model's 'scalable' functionality was enabled, allowing higher-resolution input photos to be processed within the available hardware constraints. This can sometimes reduce precision and it introduces a regular pattern in the output maps, which can, however, be effectively filtered using a discrete Fourier transform.

Figure 5(b) displays the normal map generated by the model, highlighting two problematic areas. The first issue appears in the top-left corner, where a light blue color is present instead of the expected redder hue, indicating a deviation in the surface normal along the x-direction. One possible cause could be insufficient lighting variation in this area between the pictures. The second issue appears on the right side of the endmill, where a blur is visible along the edge, possibly due to a shallow depth of field. Applying a mask to the model improves both areas; however, it reduces accuracy in the top-middle region of the endmill, where a green color, indicating the normal vector in the y-direction, is expected.

Figure 5 also shows the BRDF output of the model. The Bidirectional Reflectance Distribution Function (BRDF) describes how light interacts with an opaque surface. This model employs a Physically Based Rendering (PBR) approach using three maps: a base colour map (c), a roughness map (d), and a metallic map (e). The base colour map ($\epsilon \mathbb{R}^3$) defines the surface's primary colour. The roughness parameter ($\epsilon \mathbb{R}$) controls the sharpness or blurriness of reflections, while the metalness parameter ($\epsilon \mathbb{R}$) determines whether the surface behaves like a metal (conductive) or like a non-metal (dielectric).

The results demonstrate several advantages over the original photos. A key benefit shared by all maps is the absence of reflections, which can hinder the recognition and assessment of wear in subsequent algorithms. The normal map provides an additional advantage by containing 2.5D information, which can be processed into full 3D data if needed. This could potentially be used to estimate depth variations along the camera's axis, reducing the need for expensive 3D scanning processes. Additionally, it may enable volumetric analysis of wear forms. The base color map retains the original photo's color information, making wear particularly visible on coated cutting tools. The roughness map, however, does not appear to offer significant benefits for wear assessment. In contrast, the metallic map effectively highlights wear by revealing areas where the coating is absent. This occurs because the coating has a more metallic appearance than the exposed carbide beneath it.

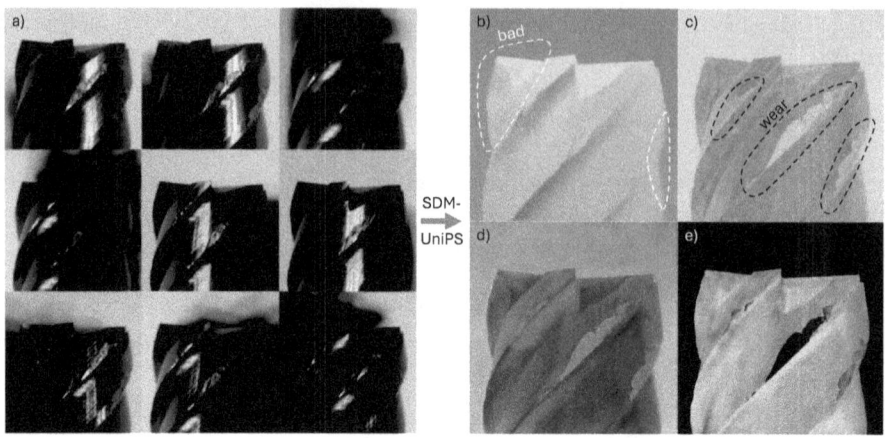

Fig. 5. Scan of an endmill: a) Input data; b) normal map; c) base colour map; d) roughness map; e) metallic map

3 Discussion

Our analysis highlights the suitability of a modern photometric stereo model for assessing tool wear in subtractive machining. While classical mathematical approaches yield insufficient results with high angular errors, the deep photometric stereo model demonstrates promising performance under the given conditions. This model operates independently of specific illumination, surface properties, or geometries, making it well-suited for analyzing various tools in the challenging environments of industrial machining. Additionally, the hardware requirements for data generation are minimal and cost-effective, enhancing its practical applicability.

However, the photometric stereo model still requires a relatively large amount of graphics memory for high-resolution images. Improving its computational efficiency is a key area for future development. While current hardware demands are reasonable for laboratory settings, real-time implementation in industrial environments will necessitate further optimization.

Future research could focus on generating optimized training datasets tailored for photometric stereo models used in microscopic tool wear observation. Additionally, the next step should be to develop new methodologies to enable automatic wear detection and assessment based on the model's output maps. Normal map integration and depth map estimation offer further potential for three-dimensional analysis. However, challenges remain due to the complex geometries of tools, particularly end mills. Achieving accurate 3D reconstructions of worn tool surfaces would be highly beneficial for detailed wear characterization and failure analysis.

In conclusion, deep photometric stereo technology shows great promise for tool wear assessment. It provides a cost-effective, non-invasive solution while overcoming many limitations of conventional imaging techniques. Future advancements in computational efficiency, automated wear detection, and 3D reconstruction will be crucial to fully harnessing its potential for industrial applications.

Acknowledgments. The project "AI-supported optimization of tool life and component quality on machine tools in machining production (KI-Span)" (original: "KI-gestützte Optimierung der Werkzeugstandzeit und der Qualität der Bauteile an Werkzeugmaschinen in der spanenden Fertigung (KI-Span)"), running from August 2021 to December 2024, is funded under the Bavarian Collaborative Research Program (BayV-FP; funding line: Digitization) by the Bavarian Ministry of Economic Affairs, Regional Development and Energy (StMWi).

Disclosure of Interests. The authors have no competing interests to declare that are relevant to the content of this article.

References

1. Zhou, Y., Xue, W.: Review of tool condition monitoring methods in milling processes. Int. J. Adv. Manuf. Technol. **96**, 2509–2523 (2018). https://doi.org/10.1007/s00170-018-1768-5
2. Shokrani, A., et al.: Sensors for in-process and on-machine monitoring of machining operations. CIRP J. Manuf. Sci. Technol. **51**, 263–292 (2024). https://doi.org/10.1016/j.cirpj.2024.05.001
3. Twardowski, P., Czyżycki, J., Felusiak-Czyryca, A., Tabaszewski, M., Wiciak-Pikuła, M.: Monitoring and forecasting of tool wear based on measurements of vibration accelerations during cast iron milling. J. Manuf. Process. **95**, 342–350 (2023). https://doi.org/10.1016/j.jmapro.2023.04.036
4. Twardowski, P., Tabaszewski, M., Pikuła, M.W., Felusiak-Czyryca, A.: Identification of tool wear using acoustic emission signal and machine learning methods. Precis. Eng. **72**, 738–744 (2021). https://doi.org/10.1016/j.precisioneng.2021.07.019
5. Yuan, J., Liu, L., Yang, Z., Bo, J., Zhang, Y.: Tool wear condition monitoring by combining spindle motor current signal analysis and machined surface image processing. Int. J. Adv. Manuf. Technol. **116**, 2697–2709 (2021). https://doi.org/10.1007/s00170-021-07366-y
6. Unal, P., Deveci, B., Ozbayoglu, M.: A review: sensors used in tool wear monitoring and prediction, pp. 193–205 (2022). https://doi.org/10.1007/978-3-031-14391-5_15
7. Munaro, R., Attanasio, A., Del Prete, A.: Tool wear monitoring with artificial intelligence methods: a review. J. Manuf. Mater. Process. **7** (2023). https://doi.org/10.3390/jmmp7040129
8. Wang, K., Wang, A., Wu, L., Xie, G.: Machine tool wear prediction technology based on multi-sensor information fusion. Sensors **24**, 2652 (2024). https://doi.org/10.3390/s24082652
9. Stephenson, D.A.: Metal Cutting Theory and Practice, 3rd edn, J. S. Agapiou, Taylor & Francis Group, Baton Rouge (2016)
10. Li, Y., Mou, W., Li, J., Liu, C., Gao, J.: An automatic and accurate method for tool wear inspection using grayscale image probability algorithm based on bayesian inference. Robot. Comput.-Integr. Manuf. **68**, 102079 (2021). https://doi.org/10.1016/j.rcim.2020.102079
11. Hu, S., et al.: Attention mechanism based CNC milling cutter wear detection using machine vision. In: 29th International Conference on Mechatronics and Machine Vision in Practice (M2VIP) (2023). https://doi.org/10.1109/M2VIP58386.2023.10413371
12. Bilal, M., Podishetti, R., Koval, L., Gaafar, M.A., Grossmann, D., Bregulla, M.: Automatized end mill wear inspection using a novel illumination unit and convolutional neural network. IEEE Access **12**, 124282–124297 (2024). https://doi.org/10.1109/ACCESS.2024.3454692
13. Brzozowski, D., Wieczorowski, M., Gapiński, B.: Geometry measurement and tool surface evaluation using a focus-variation microscope. Mechanik, pp. 1020–1022 (2017). https://doi.org/10.17814/mechanik.2017.11.167
14. Burek, J., Jamuła, B.: The accuracy of cutting edge wear measurement using a focus-variation microscope. Mechanik, pp. 850–852 (2018). https://doi.org/10.17814/mechanik.2018.10.141

15. Olortegui-Yume, J.A., Kwon, P.Y.: Crater wear evolution in multilayer coated carbides during machining using confocal microscopy. J. Manuf. Process. **9**, 47–60 (2007). https://doi.org/10.1016/S1526-6125(07)70107-X
16. Dawson, T.G., Kurfess, T.R.: Quantification of tool wear using white light interferometry and three-dimensional computational metrology. Int. J. Mach. Tools Manuf **45**, 591–596 (2005). https://doi.org/10.1016/j.ijmachtools.2004.08.022
17. Čerče, L., Pušavec, F., Kopač, J.: 3D cutting tool-wear monitoring in the process. J. Mech. Sci. Technol. **29**, 3885–3895 (2015). https://doi.org/10.1007/s12206-015-0834-2
18. Meyer, R.N., Wu, S.M.: Optical contour mapping of cutting tool crater wear. Int. J. Mach. Tool Des. Res. **6**, 153–170 (1966). https://doi.org/10.1016/0020-7357(66)90021-7
19. Ackermann, J., Goesele, M.: A survey of photometric stereo techniques. Found. Trends® Comput. Graph. Vision **9**, 149–254 (2015). https://doi.org/10.1561/0600000065
20. Ren, M., Wang, X., Xiao, G., Chen, M., Fu, L.: Fast defect inspection based on data-driven photometric stereo. IEEE Trans. Instrum. Meas. **68**, 1148–1156 (2019). https://doi.org/10.1109/TIM.2018.2858062
21. Cao, Y., et al.: Photometric-stereo-based defect detection system for metal parts. Sensors **22** (2022). https://doi.org/10.3390/s22218374
22. Saiz, F.A., Barandiaran, I., Arbelaiz, A., Graña, M.: Photometric stereo-based defect detection system for steel components manufacturing using a deep segmentation network. Sensors **22** (2022). https://doi.org/10.3390/s22030882
23. Kaya, B.: Surface roughness inspection in milling operations with photometric stereo and PNN. Int. J. Adv. Manuf. Technol. **81**, 1215–1222 (2015). https://doi.org/10.1007/s00170-015-7249-1
24. Olortegui-Yume, J., Kwon, P.: Tool wear mechanisms in machining. Int. J. Mach. Machinab. Mater. **2**, 316 (2007). https://doi.org/10.1504/IJMMM.2007.015469
25. International Organization for Standardization (ISO): ISO 8688-2; Tool Life Testing in Milling; Part 2: End Milling. Geneva, Switzerland (1989)
26. Venkatesh, V.C., Satchithanandam, M.: A discussion on tool life criteria and total failure causes. CIRP Ann. **29**, 19–22 (1980). https://doi.org/10.1016/S0007-8506(07)61288-8
27. Astakhov, V.P.: The assessment of cutting tool wear. Int. J. Mach. Tools Manuf. **44**, 637–647 (2004). https://doi.org/10.1016/j.ijmachtools.2003.11.006
28. Ikehata, S.: Scalable, Detailed and Mask-Free Universal Photometric Stereo (2023). https://doi.org/10.48550/arXiv.2303.15724
29. Ju, Y., Lam, K.-M., Xie, W., Zhou, H., Dong, J., Shi, B.: Deep Learning Methods for Calibrated Photometric Stereo and Beyond (2022). https://doi.org/10.48550/arXiv.2212.08414
30. Unsin, S., Müller, B., Schirmeier, F.: Towards real-time tool wear detection on edge devices: a lightweight dimensionality reduction approach for spindle integrated cutting force sensor data. Manuscript submitted for publication in DEXA, AI4IP (2025)

Multi-objective Reinforcement Learning for Energy-Efficient Industrial Control

Georg Schäfer[1,2,3(✉)], Raphael Seliger[4], Jakob Rehrl[1,2], Stefan Huber[1,2], and Simon Hirlaender[3]

[1] Josef Ressel Centre for Intelligent and Secure Industrial Automation, Salzburg, Austria
georg.schaefer@fh-salzburg.ac.at
[2] Salzburg University of Applied Sciences, Salzburg, Austria
[3] Paris Lodron University of Salzburg, Salzburg, Austria
[4] Kempten University of Applied Sciences, Kempten, Germany

Abstract. Industrial automation increasingly demands energy-efficient control strategies to balance performance with environmental and cost constraints. In this work, we present a multi-objective reinforcement learning (MORL) framework for energy-efficient control of the Quanser Aero 2 testbed in its one-degree-of-freedom configuration. We design a composite reward function that simultaneously penalizes tracking error and electrical power consumption.

Preliminary experiments explore the influence of varying the energy penalty weight, α, on the trade-off between pitch tracking and energy savings. Our results reveal a marked performance shift for α values between 0.0 and 0.25, with non-Pareto optimal solutions emerging at lower α values, on both the simulation and the real system. We hypothesize that these effects may be attributed to artifacts introduced by the adaptive behavior of the Adam optimizer, which could bias the learning process and favor bang-bang control strategies. Future work will focus on automating α selection through Gaussian Process-based Pareto front modeling and transitioning the approach from simulation to real-world deployment.

Keywords: Reinforcement Learning · Energy Optimization · Industrial Control · Multi-Objective Optimization

1 Introduction

Increasing energy prices and growing environmental awareness have become critical factors in industrial automation. Therefore, scientists and practitioners are trying to find new solutions to reduce energy consumption in industrial automation while still achieving high performance. This drives the need for advanced control strategies and hardware optimizations to effectively balance these competing goals [1]. This challenge is particularly evident in aerospace, where energy-efficient control strategies are crucial for battery-powered systems such as drones, ensuring extended flight durations [2].

Reinforcement Learning (RL), unlike classical methods such as Model Predictive Control (MPC) and Linear-quadratic regulator (LQR), does not require an explicit system model and can learn directly from interaction with the environment [3]. While our previous studies [3,4] focused on optimizing trajectory tracking using RL, it did not explicitly incorporate energy efficiency. In this work, we extend this approach by using Multi-Objective Reinforcement Learning (MORL), which enables the simultaneous optimization of multiple, often conflicting objectives [5] like tracking accuracy and energy minimization through energy-aware reward shaping.

1.1 Related Work

RL has shown promise in handling nonlinear dynamics, uncertainty, and partial observability in various control tasks, including UAV stabilization [6] and trajectory tracking [3,7]. However, most traditional RL algorithms focus on a single objective [8], while many industrial control problems require balancing multiple conflicting goals.

MORL addresses this by either scalarizing multiple objectives using weighted sums [9] or by directly handling vector-valued rewards via Pareto dominance to capture optimal trade-offs [10]. In parallel, recent studies have integrated energy-awareness into RL reward design through composite reward functions that balance performance metrics with energy consumption [2,11–13]. To our knowledge, however, no prior work has applied RL-based Pareto modeling to simultaneously address energy minimization and tracking accuracy in twin-rotor or similar platforms.

2 System Overview

For the experimental evaluation, we employ the Quanser Aero 2 testbed in its 1-Degree of Freedom (DoF) configuration for energy-aware control of the system's pitch angle.[1] The system is operated using two motors with a single voltage input u. The control signals are applied such that one motor receives $u_1 = u$ while the other receives $u_2 = -u$, with the input voltage bounded within – 24 V. Experiments are conducted in a Simulink-based simulation interface with Python (for further details see [14]).

The system model shown in Sect. 2 is used to predict the pitch angle φ and motor currents i of the Aero 2. The thruster-blocks contain a DC motor model that neglects armature inductance and incorporates velocity-proportional load/friction torque, described by:

$$u = iR + u_m, \quad u_m = k_m \omega_m \qquad (1)$$

$$J_m \frac{d\omega_m}{dt} = T_m - D_m \omega_m, \quad T_m = k_m i \qquad (2)$$

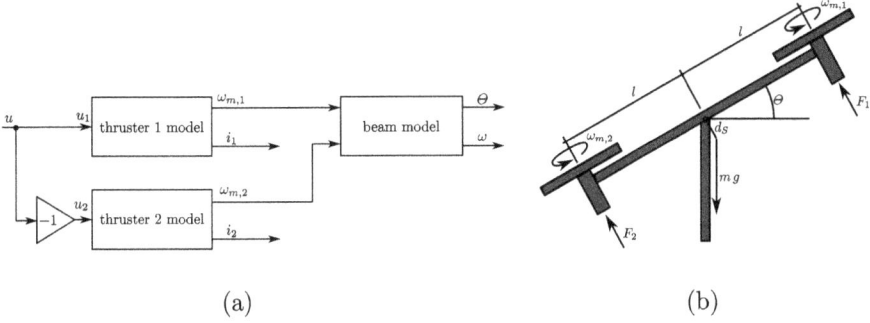

Fig. 1. (a) System model block diagram. (b) Schematic of physical testbed.

In Eq. (1) and Eq. (2), ω_m denotes the angular velocity of the thrusters, R represents the armature resistance, k_m is the motor constant, and J_m is the inertia of the motor shaft and thrusters. The back-EMF voltage and motor torque are denoted by u_m and T_m, respectively, while i represents the motor current.

The pitch model builds on the equations presented in [3], but departs from the assumption that thruster force is proportional to motor voltage. Instead, thruster force is now modeled as proportional to the thrusters angular velocity.

3 Problem Formulation

The primary control objective is to track and maintain a desired pitch angle r while minimizing energy consumption. While previous work [3,4] focused solely on tracking performance, this study extends the formulation by incorporating energy minimization. To achieve this, we define a composite reward function within our RL framework that balances tracking accuracy and energy efficiency:

$$R_t = -(1-\alpha) \cdot |\Delta_t| - \alpha \cdot P_t,$$

where $\alpha \in [0,1]$ is the energy penalty weight, Δ_t represents the normalized deviation from the target pitch, and P_t denotes the normalized electrical power consumption.

Our methodology systematically analyzes the impact of different α values on the trade-off between pitch tracking and energy savings. For the training procedure, five α values were initially evaluated with a step size of 0.25 (0.0, 0.25, 0.5, 0.75, 1.0). Each training run was conducted for 500 000 steps, with an evaluation phase every 10 000 steps, during which the model was stored for further analysis. Following the observation of a significant performance deviation when transitioning from $\alpha = 0.0$ to $\alpha = 0.25$, additional values (0.05 and 0.10) were tested. Each experiment was executed five times with different random seeds, and

---

[1] Note that in this configuration, the Aero 2 is mounted on a fixed base and cannot take off; it is designed solely to actuate and measure pitch in a single axis.

performance was evaluated based on the mean and standard deviation of pitch deviation (in degrees) and power consumption (in watts). For each value of α, a single simulation-trained agent was deployed directly on the real system without any additional training or fine-tuning, allowing us to assess its performance in a real-world setting.

4 Preliminary Results and Discussion

The results, summarized in Fig. 2, indicate a notable performance gap when adjusting the energy penalty weight. Each data point represents the mean performance over five runs, with non-Pareto solutions marked by an "×" and Pareto optimal points indicated by "o". Rectangles surrounding the markers denote the standard deviation in power consumption (watts) and tracking deviation (degrees) across the five trained agents, and colors correspond to the specific α values used. Real-system evaluations are indicated by triangles.

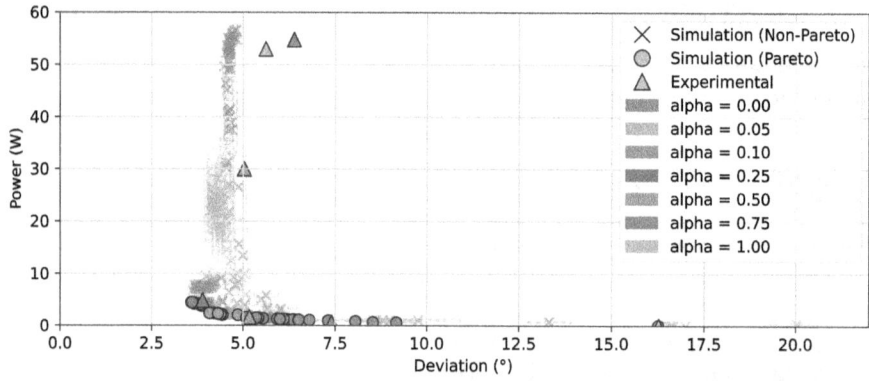

Fig. 2. Pareto front of RL solutions for energy-efficient control.

The real-world performance mirrored the simulation trends: agents trained with $\alpha \leq 0.25$ exhibited aggressive control actions, causing large overshoots and persistent oscillations. In contrast, for higher α values, the incorporation of the energy penalty served as a regularizer, resulting in smoother control responses and Pareto optimal performance. Preliminary experiments indicate that this effect is less pronounced when Adam is replaced by Stochastic Gradient Descent (SGD). These observations suggest that the aggressive behavior at low α may be partially attributed to the adaptive nature of the Adam optimizer. This highlights the need for further investigation into optimizer behavior, especially in multi-objective settings.

5 Next Steps

Our initial experiments have been conducted in simulation with manually chosen values for α. Future work aims to automate the selection of α values and further improve sample efficiency by leveraging leveraging Gaussian Process (GP) to model the Pareto front within the framework of Multi Objective Bayesian Optimization (MOBO). This approach is expected to reduce the number of required training samples by efficiently approximating the trade-offs between tracking performance and energy consumption. Additionally, future efforts will focus on transitioning from simulation-based training to real-world deployment on the Quanser Aero 2 hardware to validate the proposed methodology in a practical setting.

In parallel, we will extend these automated sampling techniques from simulation to the fixed-base Quanser Aero 2 testbed to validate the methodology in a real-world setting. Moreover, the observed non-Pareto behavior for $\alpha \leq 0.25$ raises intriguing questions regarding the utilization of adaptive optimizers like Adam in multi-objective scenarios; investigating and mitigating these artifacts will be a crucial aspect of our future research.

Acknowledgments. Financial support for this study was provided by the Christian Doppler Association (JRC ISIA), the corresponding WISS Co-project of Land Salzburg, the European Interreg Österreich-Bayern project BA0100172 AI4GREEN.

References

1. Carabin, G., Wehrle, E., Vidoni, R.: A review on energy-saving optimization methods for robotic and automatic systems. Robotics **6**(4), 39 (2017)
2. Ramezani, M., Amiri Atashgah, M.A.: Energy-aware hierarchical reinforcement learning based on the predictive energy consumption algorithm for search and rescue aerial robots in unknown environments. Drones **8**(7), 283 (2024)
3. Schäfer, G., Rehrl, J., Huber, S., et al.: Comparison of model predictive control and proximal policy optimization for a 1-DOF helicopter system (2024)
4. Schäfer, G., Krau, T., Rehrl, J., Huber, S., et al.: The crucial role of problem formulation in real-world reinforcement learning. In: 2025 IEEE 8th International Conference on Industrial Cyber-Physical Systems (ICPS) (2025)
5. Hayes, C.F., Rădulescu, R., Bargiacchi, E., et al.: A practical guide to multi-objective reinforcement learning and planning. Auton. Agent. Multi-Agent Syst. **36**(1), 26 (2022)
6. Zhang, J., Rivera, C.E.O., Tyni, K., Nguyen, S.: AirPilot: Interpretable PPO-based DRL auto-tuned nonlinear PID drone controller for robust autonomous flights (2024)
7. Jiang, Z., Lynch, A.F.: Quadrotor motion control using deep reinforcement learning. J. Unmanned Veh. Syst. **9**(4), 234–251 (2021)
8. Liu, C., Xu, X., Hu, D.: Multiobjective reinforcement learning: a comprehensive overview. IEEE Trans. Syst. Man Cybern. Syst. **45**(3), 385–398 (2014)

9. Kim, I.Y., de Weck, O.L.: Adaptive weighted sum method for multiobjective optimization: a new method for Pareto front generation. Struct. Multidisc. Optim. **31**(2), 105–116 (2006)
10. Vamplew, P., Dazeley, R., Berry, A., Issabekov, R., Dekker, E.: Empirical evaluation methods for multiobjective reinforcement learning algorithms. Mach. Learn. **84**(1), 51–80 (2011)
11. Jendoubi, I., Bouffard, F.: Multi-agent hierarchical reinforcement learning for energy management. Appl. Energy **332**, 120500 (2023)
12. Liao, Y., Friderikos, V.: Energy and age Pareto optimal trajectories in UAV-assisted wireless data collection. IEEE Trans. Veh. Technol. **71**(8), 9101–9106 (2022)
13. Yang, D., Wu, Q., Zeng, Y., Zhang, R.: Energy tradeoff in ground-to-UAV communication via trajectory design. IEEE Trans. Veh. Technol. **67**(7), 6721–6726 (2018)
14. Schäfer, G., Schirl, M., Rehrl, J., Huber, S., et al.: Python-based reinforcement learning on simulink models. In: 11th International Conference on Soft Methods in Probability and Statistics (SMPS 2024) (2024)

Deep Learning-Based Defect Detection in Laser Powder Bed Fusion

Cindy Buhl[1], Faiza Waheed[2(✉)], and Ulrich Göhner[1(✉)]

[1] University of Applied Sciences, 87435 Kempten, Germany
[2] Technical University of Applied Sciences, 83024 Rosenheim, Germany

Abstract. Additive manufacturing enables the creation of complex geometries for various applications, such as dental, medical, prototyping, and aerospace components. Despite its advantages, the printing process can encounter errors due to its complexity and numerous influencing factors, necessitating real-time anomaly detection and classification. This work introduces a non-destructive method for defect detection by monitoring layer-wise image data of the L-PBF process using a YOLOv11 object detection model to ensure component quality. A novel approach combines multiple grayscale images from a single print sequence into a 3-channel image, incorporating information from the build area in the powder bed to enhance defect classification reliability. The determination of various defect classes is also a key aspect of this work, crucial for informing subsequent interventions and decisions. Tested on a dataset generated with a Trumpf TruPrint 1000 machine, this method achieved visually promising results.

Keywords: Laser powder bed fusion (L-PBF) · Additive manufacturing · Powder bed defect detection · Deep transfer learning

1 Introduction

Laser Powder Bed Fusion (L-PBF) is a high-precision production process within additive manufacturing, playing a critical role in many industries [3]. This paper focuses on its application in metal printing.

The powder-based laser melting process of a standard L-PBF system starts by the construction of a CAD model of a metal part, transferred to a sliced representation. To minimize oxidation effects during the printing process metal powder particles are protected by a gas flow, which can be of argon or nitrogen. A powder layering device, a so-called recoater or scraper applies a new layer of powder onto the building platform. This process step is referred to as powder spreading or recoating. Subsequently the laser device melts the powder (2D slice information) along a predefined scanning path [3]. When laser energy is applied to the metal powder surface, heat is introduced, which leads to the fusion in melted zones and solidifies quickly afterwards [5,9]. After each laser scan the building platform is lowered by a distance equivalent to the layer thickness, and once the previous layer solidifies, powder spreading is repeated. These process steps are repeated until the component is finished [3].

1.1 Defect Classification and Process Influences

Due to the fact that complex physics like absorption, transmission and reflection [6] of the laser energy, adhesion of micron-scale particles, rapid melting, solidification, molten metal flow, metal evaporation and microstructural evolution are involved in the melting process, a huge variety of powder bed anomalies, which referred to as powder bed defects, can occur [14]. Other process-related and adjustable parameters, which define the building conditions, have a direct impact on the success of the build and include among others [14]: Laser power, scanning speed, scanning strategy [7], scan pattern, part geometry and orientation [11], recoater type, layer thickness, powder properties like particle size distribution, packing density, flowability, sphericity of the powder particles [2].

To ensure a comprehensive consideration of the entire process for defect classification, and due to the lack of standardized terminology stemming from the novelty of the L-PBF process, Table 1 will delve into several various defect types and their potential causes in greater detail [3,8].

Defect classes with class-ID after recoating defined in our work with relation to the introduced types are as follows, related examples are shown in Fig. 1: **0. Protruding.** Defect of rough surface with elevation of the powder bed at build areas after coating step. **1. Groove. 2. Pore.** Defect of black dots on the build area after coating step. **3. Part defect.** Significant lack of powder area in relation to the build area. **4. Incomplete Spreading.** Defect located at powder bed only areas. **5. Debris.**

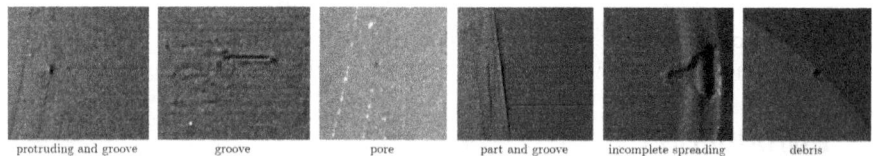

Fig. 1. Examples of labeled defect classes. Courtesy of [10].

Given the high precision required in producing metal parts, even minor defects can significantly impact the quality of the printed product, potentially compromising its mechanical strength, dimensional accuracy, surface quality, and overall performance [12].

Table 1. Defect types, causes, and visual effects.

Defect Type	Possible Cause	Visual Effect
Warpage	Support material breakage due to residual thermal stresses Insufficient welding between layers	Edge warping
Lack of Powder/Incomplete Spreading/Insufficient Powder Spreading	Powder shortage [13]	Parts can be partially covered with powder or exposed (bright areas). Continous layer defect shows height difference with bright and dark border
Recoater Streaking/Groove/Ditch	Recoater blade pulls small fragments over powder bed surface Recoater is damaged by sharp part corners	Visual scratches or grooves parallel to the recoater's movement direction
Debris	Small parts can be dragged to other powder bed locations by the recoater Broken support structures near warpage can also form debris Parts can be ejected due to small adhesion to remote powder bed areas	Dark area with recognizable edge Through the halo effect or bright edge effect, the defect can be differentiated from other defect types
Spatter	Molten metal droplets and unmelted powders are ejected from the melt pool often above components against the gas flow direction	Small black particles
Porosity	Non-well melted areas during the melting stage [4]	Black circular dots
Super-elevation/Protruding	Residual thermal stresses [1], layer thickness, material deposition or swelling	Part wraps or curls upwards of the powder layer [1]

2 Literature Review

Zhao et al. [13] utilized three different deep learning object detection models for defect detection in L-PBF with low-resolution images, defining only three types of defects. The YOLOv5x model demonstrated the best results, achieving over 92% accuracy for each detection class. Scime et al. [8] proposed a resource-intensive pixel-wise localization using the DSCNN model for L-PBF anomalies, identifying 12 different surface-visible anomalies. They stacked post-spreading and calibrated post-fusion images into two channels, achieving a validation specificity ranging from 99.9% to 25.1%.

3 Methods and Data

From [10] over 20,188 images of Stainless Steel 316L parts, not yet publicly available, recorded with a Basler acA3800–14um camera and a Basler 4 mm Lens C125–0418–5M-P were provided. Each print sequence contains of 4 images that are related to different but fixed image capture settings, whereas only the image of post-fusion defined as exposed slice and 2 images of post-spreading with different lightning conditions, are considered. In this study only a total of 224 images, due to the small defect instances and a lack of sufficient time and ressources, were labeled with 6 defect classes, corresponding to 56 print sequences. Defect classes are divided into 530 protruding, 156 groove, 190 pore, 37 part, 46 incomplete spreading and 4 debris defect instances. For training, we stacked 3 out of the 4 sequence images (exposed slice, powder bed light right, powder bed light left) as a synthetical 3-channel input. The powder bed with top light was excluded, due to the fact that it does not add additional information.

We treated the 3 images of each print sequence as a single process, ensuring defects were consistently labeled across images. Identical object instances appearing in multiple images of the same sequence where only considered once.

We trained a COCO-pretrained YOLOv11m model for 1000 epochs with early stopping, using the highest available batch size of 4. Training was conducted on an NVIDIA GeForce RTX 2080 Ti with 11 GB of memory, an input size of 1088x1088x3 and a train-validation split of 0.75–0.25.

4 Preliminary Results and Conclusions

A comprehensive literature review identified potential defect classes and process influences. Although training a YOLOv11 model showed promising prediction, as depicted in Fig. 2. There is still room for improvement. The model operates with an affordable camera system and achieves an inference time of just a few milliseconds. By considering multiple grayscale images from a single print sequence, it accounts defect class dependencies on the part area and incorporates pre-researched defect classes and process influences. Limitations include class imbalance and poor validation results.

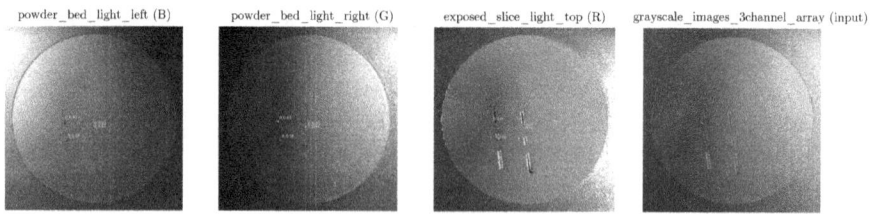

Fig. 2. Prediction results on an unseen sample visualized by channel plus a 3-channel array representation. Underlying image data originates from [10].

5 Next Steps

While the current results are promising, quantitative evaluation, applying model optimizations and conducting model benchmarking are necessary.

Future work will focus on additional labeling, data augmentation and the incorporation of diverse materials and geometries to enhance model resilience. The results can then be utilized to trigger actions such as sending warnings or adjusting process parameters, which can contribute to a sustainable manufacturing process by reducing waste and increasing energy efficiency.

Acknowledgments. Financial support for this study was provided by the European Interreg Österreich-Bayern project BA0100172 AI4GREEN.

Disclosure of Interests. The authors have no competing interests to declare that are relevant to the content of this article.

References

1. Abdelrahman, M., Reutzel, E.W., Nassar, A.R., Starr, T.L.: Flaw detection in powder bed fusion using optical imaging. Addit. Manuf. **15**, 1–11 (2017)
2. Boschetto, A., Bottini, L., Vatanparast, S.: Powder bed monitoring via digital image analysis in additive manufacturing. J. Intell. Manuf. **35**(3), 991–1011 (2024)
3. Cheng, L., et al.: Low-rank adaptive transfer learning based for multi-label defect detection in laser powder bed fusion. Opt. Lasers Eng. **184**, 108683 (2025)
4. Dinh, D., Muller, N., Quinsat, Y.: Layering defects detection in laser powder bed fusion using embedded vision system. Comput.-Aided Des. Appl. **18**(5), 1111–1118 (2021)
5. Grasso, M., Colosimo, B.M.: Process defects and in situ monitoring methods in metal powder bed fusion: a review. Meas. Sci. Technol. **28**(4), 044005 (2017)
6. Pandiyan, V., et al.: Deep transfer learning of additive manufacturing mechanisms across materials in metal-based laser powder bed fusion process. J. Mater. Process. Technol. **303**, 117531 (2022)
7. Rodriguez, E., et al.: Integration of a thermal imaging feedback control system in electron beam melting. In: 23rd Annual International Solid Freeform Fabrication Symposium - An Additive Manufacturing Conference, SFF 2012, pp. 945–961 (2012)
8. Scime, L., Siddel, D., Baird, S., Paquit, V.: Layer-wise anomaly detection and classification for powder bed additive manufacturing processes: a machine-agnostic algorithm for real-time pixel-wise semantic segmentation. Addit. Manuf. **36**, 101453 (2020)
9. Sharratt, B.M.: Non-destructive techniques and technologies for qualification of additive manufactured parts and processes: a literature review: Report (2015)
10. Waheed, F.: Laser powder bed images of a Trumpf TruPrint1000 (2025)
11. Wohlers, T., Campbell, R.I., Diegel, O., Kowen, J., Mostow, N.: Wohlers report 2021: 3D Printing and Additive Manufacturing: Global State of the Industry. Wohlers Associates, Fort Collins (2021)
12. Zhao, J., et al.: Real-time detection of powder bed defects in laser powder bed fusion using deep learning on 3D point clouds. Virtual Phys. Prototyp. **20**(1) (2025)

13. Zhao, Y., Ren, H., Zhang, Y., Wang, C., Long, Y.: Layer-wise multi-defect detection for laser powder bed fusion using deep learning algorithm with visual explanation. Optics Laser Technol. **174**, 110648 (2024)
14. Zitelli, C., Folgarait, P., Di Schino, A.: Laser powder bed fusion of stainless steel grades: a review. Metals **9**(7), 731 (2019)

Prediction of CNC Manufacturing Time Under Real-World Conditions Using Graph Convolutional Networks

Fabio Lischka[✉], Andreas Schwarz, Dominik Wiesner, Christoph Wald, Frank Schirmeier, and Ulrich Göhner

Institute for Data-Optimised Manufacturing, UAS Kempten, Bahnhofstr. 61, Kempten, Germany
{fabio.lischka,andreas.schwarz,dominik.wiesner,christoph.wald,
frank.schirmeier,ulrich.goehner}@hs-kempten.de

Abstract. In this work, we share our learnings on predicting CNC manufacturing time given CAD models from a real-world dataset using machine learning models. To minimize prediction cost, we focus on extracting relevant information solely from the CAD model, eliminating the need for manual feature labeling. Our experiments reveal that a combination of hand-crafted features with those automatically extracted by the graph convolutional network UV-Net yields the most accurate predictions. Notably, our model exhibits robust performance when applied to a newer, higher-quality dataset, achieving a significant improvement of 32% in mean absolute error compared to a rule-based approach despite the challenges posed by temporal and data quality shifts.

Keywords: CNC machining · Machining time prediction · Machine Learning · Deep Learning

1 Introduction

Computerized Numerical Control (CNC) machining is a widely adopted technique for manufacturing mechanical parts from various materials, including steel, aluminum, and plastic. Its flexibility enables efficient production of small batches with similar geometries. However, generating the Numerical Control (NC) code that specifies machine operations via CAM software (Computer Aided Machining) can be time-consuming especially when producing small batches, so the NC code may not be available until shortly before production begins.

Accurate estimation of CNC manufacturing time is crucial for both production planning and cost estimation. Once the NC code has been generated, such an estimate can be obtained by simulating the execution of the NC code. Moreover, multiple publications have reported significant improvements in estimating machining times using rule-based corrections [8] and/or machine learning models [9]. However, these approaches rely on the availability of the NC code, which may not be available at the time of planning or cost estimation.

To address this challenge, we propose a deep learning model that estimates machining time prior to NC code generation, using only the 3D CAD model (Computer Aided Design) in the STEP format (Standard for the Exchange of Product model data) as input. Compared to traditional rule-based estimates, this approach provides more accurate predictions at an early stage, enabling better production planning and resource allocation. Unlike many rule-based estimates, our approach does not rely on hand-labeled features as input for prediction, e.g. information about machining features. Employing human experts for manual labeling is time-consuming, cost-intensive and error-prone [1]. By avoiding the associated costs and delays, our proposed model offers significant benefits, particularly for the production of small batches.

Our proposed model is trained on a large, contemporary dataset of 6419 real-world machined parts contributed by a globally operating manufacturing company. Our dataset contains diverse geometries and includes real machining times as labels. Albeit collected on a single machine model, these machining times have been collected over an extended time span and thus cover a variety of real-world sources of variation, including in particular the influences of the NC code creator. Our contribution is to demonstrate the advantages of machine learning models, particularly graph neural networks, over rule-based approaches in this real-world setting.

2 Prior and Related Work

Our work extends the existing literature on this topic, which has been limited by several factors. Previous studies have relied on manual annotations of machining features or part complexity as input for their machine learning models [1,4,6], or have focused on specific domains, particularly plastic mold forms, leaving it unclear whether the results generalize to CNC parts from other domains [3,4,6].

3 Dataset

Our main dataset comprises 6419 parts machined on a single model of modern five-axis CNC machine over a fixed time period. This diverse collection includes parts from more than 20 different classes, resulting in significant variations within the dataset. For each part in the dataset, the following information is available:

- the CAD file in STEP format (a representation of the geometry of the part with no further annotation);
- a proprietary, rule-based time estimate derived from Enterprise Resource Planning (ERP) and CAD data, serving as our baseline for comparison;
- the simulated machining time predicted by CAM software (not available at planning time);
- the actual machining time recorded by the CNC machine control, serving as our ground truth.

To ensure data quality, we applied filters to remove parts with machining times under 300 s or above 20,000 s, as these likely indicate interrupted or abnormal machining processes. Additionally, we excluded part classes with fewer than five associated parts to guarantee sufficient representation for meaningful machine learning generalization.

To obtain a conservative estimate of our chosen machine learning model's performance under production conditions, we utilize a separate test dataset comprising 642 parts manufactured during a later time frame. This test dataset contains the same annotations as the main dataset and underwent additional filtering to remove aborted machining runs and other irregularities. These filters could not be applied to the main dataset because the filtering relies on additional information not available for the main dataset. While this higher data quality may have a negative effect on test performance of models trained on the main dataset, this filtering is crucial to evaluate our trained models against existing estimators.

4 Machine Learning Architectures Used

Previous research on estimating machining times primarily employed shallow neural networks with simple hand-crafted features and information about machining features as input [1,3,4,6]. However, since information about machining features is not available in our use case, we choose to combine hand-crafted features with features extracted from the STEP file using deep neural networks.

We compute a set of hand-crafted geometric features from the STEP files, including a subset of the features described in [4], as well as counts of faces and vertices. We filter these hand-crafted features after considering their correlations with the machining time as well as with each other to avoid adding irrelevant or redundant input features.

As a baseline approach, we train a shallow artificial neural network with a single hidden layer on seven of the hand-crafted features, following [3,4]. We compare this approach with two deep learning architectures: a 3D convolutional neural network (CNN) that receives a voxel representation of the CAD model as input [7], and UV-Net [5], a graph convolutional neural network that operates on an enriched graph representing the CAD model. A third architecture, Hierarchical CADNet [2], failed to converge on our training dataset.

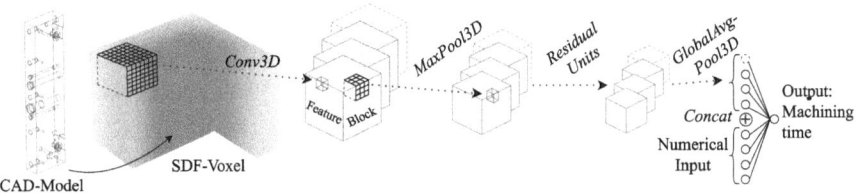

Fig. 1. The architecture of our Voxel CNN.

Our *Voxel CNN* employs the Signed Distance Function (SDF) for representing inputs, so each voxel contains the signed distance to the surface of the machined part, i.e. the negative distance inside and positive distance outside of the part. The input is a 100×100×100 voxel grid, providing a high-resolution volumetric representation of the shape. Each filter in the convolutional layers generates a new feature block, facilitating hierarchical feature extraction. The architecture extends the concept of residual networks to 3D convolutional domains, using residual units comprising two 3D convolutional layers with batch normalization and an activation function. The architecture is shown in Fig. 1.

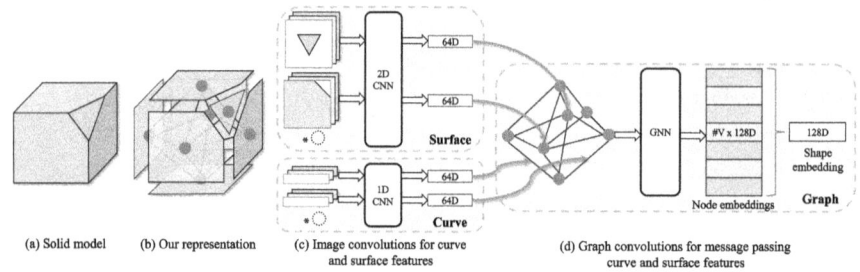

Fig. 2. The architecture of UV-Net. Graphic from [5].

UV-Net is a neural network architecture specifically designed for CAD representation learning, capable of directly processing the Boundary Representation (B-rep) of 3D models. For the task addressed in this paper, the network architecture was expanded to allow for regression problems to be solved. In order to capture the topological information of the 3D models, UV-Net derives the face-adjacency graph $G = (V, E)$ from the boundary representation of the CAD model, where vertices V represent faces and the edges E indicate connections between these faces. Additionally, a set of geometric features is computed from the geometry of edges and faces by 1D and 2D CNNs with learnable weights. The features are assigned as node and edge attributes to the graph G and propagated through G using graph convolution operations. The architecture is shown in Fig. 2.

To combine the strengths of both the shallow neural network and the deep learning approaches, we suggest a hybrid approach that uses both hand-crafted features as described at the beginning of this section and features generated using one of the two aforementioned deep learning architectures, see Fig. 3. The weights of the deep neural network and the prediction head are learned on the training set, while the computation of hand-crafted features remains fixed.

5 Experimental Results

We use 80% of our main dataset for training and 20% for validation. We compare the architectures described in Sect. 4:

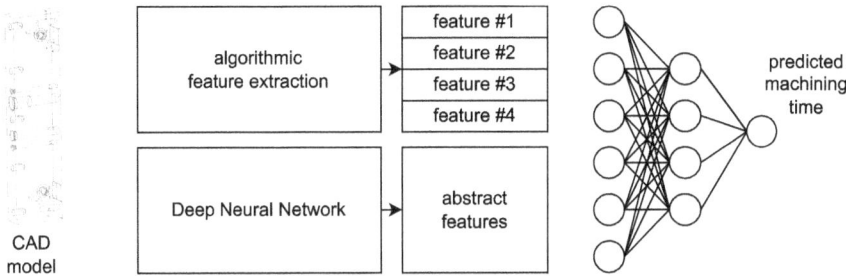

Fig. 3. The proposed hybrid architecture.

- a shallow artificial neural network (ANN) with hand-crafted features as input,
- deep neural networks: Voxel CNN and UV-Net with the CAD model as input,
- a hybrid architecture that combines either Voxel CNN or UV-Net with hand-crafted features.

We compare our results against a proprietary rule-based prediction that uses input features that are provided manually for each part (some of them hand-labeled), with a different algorithm for each class of machined parts. Furthermore, we compare against the time predicted by a simulation of the NC code (for context only, as NC code is not available at planning time). All models are initialized with random weights and trained on the training split. We use Bayesian optimization for hyperparameter tuning on the validation split, except for the shallow ANN, which uses fixed hyperparameters. Each deep learning model is trained on Nvidia H100 GPUs for 150 epochs using the Adam optimizer. As evaluation metrics, we rely mainly on MAE (mean absolute error; in seconds) and – because the actual machining times span several orders of magnitude – MAPE (mean absolute percentage error). Our results are summarized in Table 1. The rule-based prediction is only available for 343 out of 1282 machined parts in the validation split, the NC simulation for 1043 machined parts. The performance of our models is evaluated on the whole validation split.

To obtain a conservative estimate of our model's performance when deployed in production, we evaluate our best-performing model (with respect to MAPE) on the test dataset from a later time frame. We expect deteriorated performance due to two factors: data drift and better filtering of the test dataset. This gives a conservative estimate of the model performance in production, prior to mitigation strategies such as drift detection or retraining on additional filtered data. The performance on the test dataset is displayed in Table 2. The rule-based prediction is available for 588 out of 642 parts in the test dataset.

Table 1. Performance on the validation split.

	# Parameters	MAPE (↓)	MAE (↓)	R^2(↑)
NC simulation	-	0.12	485	0.94
State-of-the-art methods				
Rule-based	-	1.43	3432	-3.85
Shallow ANN	120	0.59	1735	0.34
Voxel CNN	1.7M	0.95	5917	-0.53
UV-Net	2.1M	0.50	1512	0.41
Hybrid architectures				
Hybrid using Voxel CNN	4.2M	0.74	2866	0.19
Hybrid using UV-Net	1.3M	**0.46**	**1285**	**0.55**

Table 2. Perfomance on the test dataset.

	MAPE	MAE	R^2
Rule-based	0.45	2444	**0.44**
Hybrid using UV-Net	**0.19**	**1656**	0.25
NC simulation	0.10	531	0.97

6 Discussion and Conclusion

Our results demonstrate the effectiveness of deep neural networks in predicting machining time from STEP files. A hybrid approach, combining hand-crafted with automatic feature extraction through neural networks, proves successful and may improve further if additional, even more expressive hand-crafted features are added.

The graph convolutional neural network UV-Net demonstrates strong performance, both as a standalone model and as part of a hybrid architecture. This points to the benefits of the adjacency graph representation - namely, that it provides sensible aggregations of large homogeneous faces in the CAD model. In contrast, the voxel representation has the benefit of natively describing the spacial structure of the CAD model. However, the fixed number of voxels makes it hard to capture important details like small radii and holes. Also, the voxel representation lacks explicit information about the boundaries between faces, which might complicate the extraction of relevant geometric information for the CNN. This provides a speculative explanation for the superior performance of the graph neural network.

When comparing results between validation split and test dataset, we observe that the mean absolute error of the rule-based prediction diminishes significantly, while the MAPE values of all predictors improve. This is likely the effect of irregular machine runs in the main dataset, which have been filtered out in the test dataset: Predictions might be orders of magnitude higher than the measured time of an aborted run, which has a large impact on the mean absolute percentage error.

Consequently, we expect the performance of our machine learning model to deteriorate (in terms of mean absolute error) on the higher-quality test dataset. As a large proportion of the increased mean absolute error could be attributed to this change in data quality, we expect that fine-tuning our machine learning model on high-quality data could lead to substantial improvements.

However, there may be limitations to the potential performance gains that can be achieved in this way. The lack of information about complex features like threads in STEP files means that the same CAD model can represent different parts with vastly different machining features and, hence, machining times; incorporating 2D drawings or other data sources could help address this limitation. Moreover, the variance in the machining times even for very similar parts (analysis not included) due to NC code and machine operation variations may be the limiting factor in the end. The high overall concordance of actual and NC simulation times indicates that the NC code generation constitutes the main source of variance.

Despite these limitations, our evaluation shows that machine learning approaches outperform rule-based predictions, even under changes in data distribution. This makes such models highly practical for rough planning of CNC machining production, offering more accurate predictions and eliminating the need for manual CAD labeling. Our results demonstrate the potential of machine learning to improve planning effectiveness and reduce costs for estimation.

Disclosure of Interests. The authors have no competing interests to declare that are relevant to the content of this article.

References

1. Atia, M., Khalil, J., Mokhtar, M.: A cost estimation model for machining operations; an ANN parametric approach. J. Al-Azhar Univ. Eng. Sector **12**(44), 878–885 (2017). https://doi.org/10.21608/auej.2017.19195
2. Colligan, A.R., Robinson, T.T., Nolan, D.C., Hua, Y., Cao, W.: Hierarchical CAD-Net: learning from B-reps for machining feature recognition. Comput. Aided Des. **147**, 103226 (2022). https://doi.org/10.1016/j.cad.2022.103226
3. Eraslan, E.: The estimation of product standard time by artificial neural networks in the molding industry. Math. Probl. Eng. **2009**(1) (2009). https://doi.org/10.1155/2009/527452
4. Florjanič, B., Kuzman, K.: Estimation of time for manufacturing of injection moulds using artificial neural networks-based model. Polimeri **33**(1), 12–21 (2012)
5. Jayaraman, P.K., et al.: UV-net: learning from boundary representations. In: 2021 IEEE/CVF Conference on Computer Vision and Pattern Recognition (CVPR), pp. 11698–11707. IEEE (2021). https://doi.org/10.1109/cvpr46437.2021.01153
6. Rodrigues, A., Silva, F.J.G., Sousa, V.F.C., Pinto, A.G., Ferreira, L.P., Pereira, T.: Using an artificial neural network approach to predict machining time. Metals **12**(10), 1709 (2022). https://doi.org/10.3390/met12101709

7. Wald, C., Jung, T., Schirmeier, F.: 3D convolutional neural network to predict the energy consumption of milling processes. In: Proceedings of the 14th International Conference on Data Science, Technology and Applications (2025)
8. Ward, R., Sencer, B., Jones, B., Ozturk, E.: Accurate prediction of machining fedrate and cycle times considering interpolator dynamics. Int. J. Adv. Manuf. Technol. 417–438 (2021). https://doi.org/10.1007/s00170-021-07211-2
9. Yamamoto, Y., Aoyama, H., Sano, N.: Development of accurate estimation method of machining time in consideration of characteristics of machine tool. J. Adv. Mech. Des. Syst. Manuf. **11**(4), JAMDSM0049 (2017). https://doi.org/10.1299/jamdsm.2017jamdsm0049

A Vision-Guided Approach to Pick-and-Place Robotics: From Assembly Drawings to Industrial Assembly Automation

Raphael Seliger[✉], Matthias Micheler, Sebnem Gül-Ficici, and Ulrich Göhner

Kempten University of Applied Sciences, Kempten, Germany
raphael.seliger@hs-kempten.de

Abstract. Vision-guided robotic systems play a key role in industrial automation, particularly in flexible pick-and-place tasks. We present a modular approach that integrates automatic data extraction of 2D assembly drawings, object recognition, automatic robotic part placement, and augmented reality for quality inspection. The system combines a YOLOv8-based segmentation pipeline with classical shape-matching techniques to detect, identify, and align components during assembly. We apply a hybrid method to interpret assembly drawings that merge traditional computer vision with neural segmentation masks. The robotic setup features a 6-DOF arm controlled via Robot Operating System (ROS), with overhead cameras for object localization. This paper presents ongoing work. Preliminary results indicate improved segmentation accuracy using AI-based methods for extracting the foundational data from the 2D assembly drawings. Future work will focus on enhancing grasp and path planning, increasing placement precision, and improving the robustness of part detection. Additionally, we aim to generalize the system to previously unseen assembly plans and expand the feature extraction capabilities for complex assembly drawings.

Keywords: Computer Vision · Pick-and-Place · Robotics · Industrial Automation · 2D Assembly Drawing

1 Introduction

Robotics has developed rapidly in recent years, enabling robots to handle repetitive pick-and-place tasks on production lines. [1]. However, traditional systems often need reconfiguration for new product variants, limiting their adaptability. To improve flexibility, researchers are increasingly integrating Artificial Intelligence (AI) and Computer Vision (CV) techniques into robotic operations [2–4]. Integrating AI-based perception with robot control requires various foundational components to function effectively. Our previous research identified and developed several of these components (see Fig. 1). We introduced the "Visual

Recognition Module" that merges deep learning with traditional contour-based methods to detect and localize assembly parts based on visual input [5]. We also developed the "Assembly Projection Module" capable of projecting CAD-based assembly drawings in true scale to support manual placement, enable visual comparison, and assist with quality control. In addition, we developed the "Feature Extraction Module", which analyzes heterogeneous 2D assembly drawings (including scanned or unstructured diagrams) to identify individual components and extract geometric features for downstream matching and planning. This component was the focus of our second paper, which addressed the challenge of automatically populating the reference database from a large and diverse set of assembly drawings [6]. These modules are supported by a "Reference Database" containing predefined geometric part descriptors, an "Integration Platform" for real-time coordination of all system components, and a "User Interface" that enables operators to manage diagrams and monitor system behavior [5].

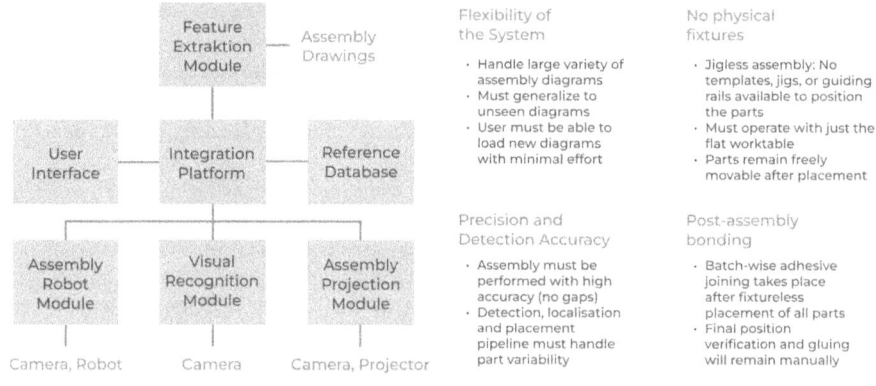

Fig. 1. This figure presents the main components of the system, combining elements from prior research with newly developed and revised components, as well as the corresponding system requirements.

This paper presents ongoing work integrating previously developed components into a unified, vision-guided robotic system capable of performing flexible, puzzle-like assembly tasks. The primary goal is to evaluate whether automatically extracted visual and geometric information from 2D assembly drawings is sufficient to enable robotic automation of a formerly manual, high-precision assembly process. To address this objective, the current work introduces the "Assembly Robot Module" that addresses the final step of the process: translating drawing-based spatial knowledge into real-world robotic motion.

1.1 Related Work

Recent research has focused on CAD-to-robot workflows that use CAD models for planning, perception, and control in pick-and-place tasks. Grasp and motion planning are commonly studied separately due to their complexity [7,8]. One system treats the CAD model as a digital twin, enabling pre-computation of grasp poses and motions. This CAD-driven planning integrates with ROS tools like MoveIt! to execute movements while considering potential collisions [9].

Synthetic training data generated from CAD models has been used to train object pose estimators and grasp detection algorithms without requiring physical components [10]. CAD-based approaches have also enabled estimation of an object's 6-DoF pose in RGB images, supporting detection of unknown objects [11]. A recent framework allows robots to pick, re-grasp, and place novel objects using CAD models, achieving over 90% placement success and improving planning flexibility, recognition accuracy, and automation integration [12].

The application scenario of this paper resembles solving Tangram or jigsaw puzzles, requiring recognition of flat geometric parts, orientation estimation, and precise placement. One approach detects the shape and pose of Tangram pieces to match predefined patterns, emphasizing the importance of accurate contour detection to prevent misalignment [13]. Other methods solve jigsaw puzzles using visual features for piece matching, typically assuming uniform shapes and fixed orientations [14]. In contrast, Tangram tasks involve ambiguous arrangements without semantic guidance. Generative models and search-based strategies have been proposed to handle the free rotation and complex configurations of polygonal pieces [15]. These studies underscore the role of strategic planning, where search algorithms determine optimal sequences of actions [13].

While most robotic assembly approaches in the literature are based on structured 3D CAD data, 2D assembly drawings remain largely unexplored as a data source despite their continued relevance in industrial practice. These drawings are often optimized for human interpretation, lack explicit geometric semantics, and are frequently available only in rasterized formats (e.g., scanned PDFs or images), which limits their direct usability in automated workflows. As a result, using computer vision to extract actionable data from such drawings introduces unique challenges for flexible robotic assembly, especially in scenarios without mechanical fixtures or interlocking parts, as described in this paper.

2 Approach and System Components

In order to investigate whether the data extracted from 2D assembly drawings is sufficient to automate a complex manual assembly process fully, several key modules need to be improved. While earlier work focused on assistive systems for assembly operations and partial automation [5,6], this paper aims to extend automation to the physical execution of assembly movements. This shift revealed new practical challenges that demanded enhancements: the "Visual Recognition Module" must be improved to increase accuracy and robustness in challenging conditions; the "Feature Extraction Module" must be refined to reliably interpret technical drawings, even when they are noisy, incomplete, or unstructured; and a new robotic module called "Assembly Robot Module" is introduced to transform reference data into Cartesian coordinates and to execute the physical pick-and-place operations.

2.1 Improvements to the Visual Recognition Module

The "Visual Recognition Module" from our previous work [5] served as the foundation for the object identification pipeline in this paper. The original implementation relied on a YOLOv8-based segmentation model to isolate each part from the background and to segment the visible surface of assembly parts, which are characterized by simple geometric shapes. Part identification was then performed using a contour-matching approach based on Hu moments, achieving an identification accuracy of 96%. However, due to the limited number and low variability of test objects, these results could not be reliably generalized to the full range of real-world assembly parts.

To address this, the present work expanded the dataset, increasing the number of unique test parts from 500 to 10.000 images (from only one to 25 unique assembly drawings). When evaluated on this broader dataset, the original pipeline's accuracy dropped to 65%, highlighting its limited robustness under more realistic conditions. This motivated several enhancements to the pipeline. First, the post-processing of the masks was improved to reduce segmentation noise and artifacts. In particular, morphological operations and shape-based filtering were applied to refine the contours before comparison. Second, additional shape descriptors were integrated alongside Hu moments to improve classification robustness. These included contour area, aspect ratio, and polygonal approximation features. Moreover, the YOLOv8 model was fine-tuned with an extended and more diverse training dataset, which included images under different lighting conditions and backgrounds to improve generalization.

2.2 Improvements to the Feature Extraction Module

This work aims to accurately detect assembly part geometries and correctly assign component IDs based on 2D assembly drawings. Each assembly part is defined by a unique geometric shape and an associated identifier, with some IDs linked to specific components via assignment arrows and others directly embedded as textual labels within the individual parts inside the assembly diagram. Key structural elements must be extracted and stored in the "Reference Database" to support automated matching tasks. Previous work explored a complex pipeline that combined the Segment Anything foundation model for initial mask extraction, ResNet-18 for component classification, TesseractOCR for text extraction, and Faster R-CNN for associating part IDs with corresponding assembly parts [6]. However, this approach proved inefficient and computationally expensive, with limited reliability when applied to highly variable assembly drawings, especially regarding unreliable ID localization due to OCR-based ID extraction errors and resolution-dependent failures in detecting arrow marks and anchor points. To address these limitations, we propose and compare two alternative methods in this paper, focusing on performance and accuracy. The first approach relies entirely on classical computer vision techniques, using rule-based segmentation and heuristic matching. The second approach leverages a more efficient AI-based pipeline, employing lightweight neural models for component detection and ID assignment, in contrast to the more complex and resource-intensive architecture presented in our earlier work [6].

Three additional OCR systems were evaluated across multiple image resolutions: Idefics, DocOwl, and EasyOCR [16–18]. EasyOCR demonstrated the highest accuracy and robustness for our specific use case. A voting matrix was implemented to consolidate ID candidates across resolutions further to enhance recognition reliability, reinforcing consistently detected text while filtering out resolution-dependent artifacts. The detection of circular anchor points (used for arrow-based ID assignments) was also improved. Hough Circle Detection was applied, followed by an algorithm that determined the direction of each arrow and interpolated along its trajectory until a corresponding OCR-identified ID was found. Finally, connected component labeling (commonly referred to as Blob Coloring) was applied to segment individual assembly parts within the assembly drawings to identify distinct regions based on contour closure. A further challenge emerged in interpreting part boundaries due to inconsistent stroke widths in CAD plans. This was addressed by defining the part boundary as the centerline between stroke edges and localizing it with subpixel precision using a filter-based method.

After presenting the classical algorithmic variant, the following section introduces the alternative AI-based pipeline. The segmentation and classification were implemented using a segmentation-focused YOLOv8 model trained on Components, ID bounding boxes, and circles of arrows to enhance segmentation. A key challenge in training the model was differentiating adjacent part segments, similar to issues in traffic scene understanding, where adjacent objects of the same class are difficult to separate. This was addressed with pre- and post-processing techniques. The image processing algorithm limited segment sizes, creating visual gaps in the segmentation mask, which aided the model in learning distinct instances. After inference, the predicted segments were expanded to align with the original contours. The solutions developed for the previously described algorithmic approach were reused without modification for the remaining sub-tasks besides the segmentation.

2.3 Introduction of the Assembly Robot Module

The "Assembly Robot Module" is responsible for physically executing the placement of components at precisely defined positions, as derived from the assembly drawing and vision-based localization. It translates the extracted data into concrete robot motions, completing the automation pipeline from perception to action. While the control logic is deliberately kept simple in the current prototype, it establishes a functional baseline for evaluating the feasibility and precision of a fully vision-guided robotic assembly.

The experimental setup includes a 6-DOF robotic arm in front of a working table, where flat, puzzle-like components are placed. The robot is equipped with a pneumatic suction gripper suitable for planar parts. Two overhead cameras monitor the pickup and placement zones, enabling vision-based localization and feedback. All components are integrated via the ROS middleware for modular communication and control. ArUco markers placed in both the pickup and placement areas serve as fixed reference points. These markers are detected during runtime to compute the homography for coordinate transformation between the camera frame and the robot workspace. Grasp points are determined based on the geometry of each component. While the center of mass derived from the outer contour is sufficient for many solid parts, it fails for parts containing internal holes, where suction may engage voids rather than material. An Euclidean distance transform is applied to the binary object mask to address this, identifying the point furthest from any contour boundary. This yields a robust and stable suction point even for complex shapes. The robot executes a pick-and-place trajectory once the target pose and grasp point are determined. A basic motion planning strategy is used: the robot moves to an intermediate approach position above the object, descends vertically for pickup, lifts the part, and performs a mirrored sequence for placement. Collision avoidance and joint-space constraints are handled via standard ROS motion planning components.

3 Preliminary Results and Discussion

While the development of individual components and the integration of the overall system are still ongoing, several preliminary results have already been obtained. The current version of the "Visual Recognition Module" achieved an accuracy of over 90% in the first controlled experiments. The grasp point estimation in the "Assembly Robot Module", based on geometric analysis and distance transforms, has proven reliable in these settings, and the coordinate transformation from the assembly diagram to the robot workspace is functioning as intended. However, accurate placement of the components remains a challenge. Errors in the placement phase frequently lead to either small gaps between parts or slight overlaps.

In parallel, two approaches for building the assembly plan annotation dataset were evaluated for the "Feature Extraction Module". The classical image processing pipeline (relying on techniques such as blob coloring) performed adequately on simple plans, achieving a mAP@50 of 0.45 across a test set of 25 CAD drawings. Specific challenges became apparent during processing, especially with components containing hatched patterns at varying angles and crosses marking borehole centers. These elements disrupted the detection of closed shapes, and the image-processing algorithm failed to handle them reliably. Filtering attempts often removed essential structural lines, particularly at the tile boundaries.

To address these limitations, a machine learning-based system was introduced to improve robustness and accurately detect such complex tile structures. The same dataset processed using a YOLOv8-based segmentation model yielded a substantially higher mAP@50 of 0.91, demonstrating the improved robustness and generalization of AI-based methods, even when trained on relatively few examples. Several components still require refinement before a final evaluation can be conducted. In particular, the placement logic must be stabilized, and the tile segmentation system must be further validated on more diverse CAD layouts.

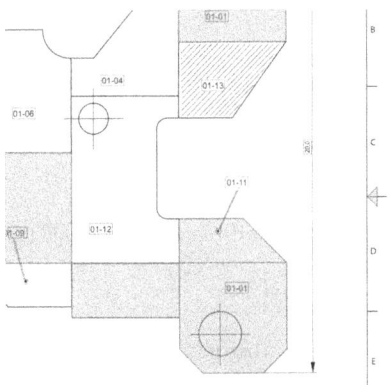 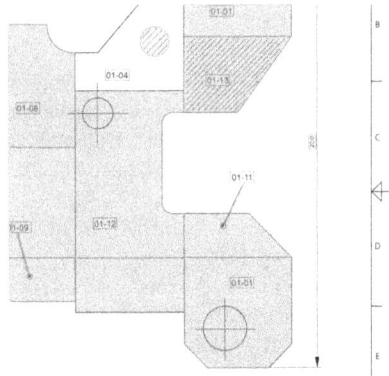

Fig. 2. Algorithmic segmentation: the components not highlighted in blue were not recognized, e.q. due to hatched parts or marked bore holes. (Color figure online)

Fig. 3. YOLO-based segmentation: Some hatched parts remain challenging, one error is highlighted in orange; improved overall accuracy. (Color figure online)

Once these adjustments are complete, a comprehensive system-level evaluation will follow to assess overall accuracy, reliability, and time performance in real-world assembly scenarios.

4 Conclusion and Future Work

Future work in the "Visual Recognition Module" will enhance the object recognition pipeline by adding more geometric and visual features utilizing DIPlib for improved low-level analysis. Additionally, we will refine rotation estimation in the "Assembly Robot Module" using template alignment or learning-based predictors. To improve placement accuracy, we will implement a strategy to adjust component positions while minimizing gaps and preventing unintended displacement. This suggests a more refined placement strategy and a revised sensing setup. Although two static overhead cameras were used in the current prototype, future versions would benefit from an eye-in-hand configuration or an additional depth-sensing unit to improve local precision during placement. In the context of automated segmentation in the "Feature Extraction Module", it was observed that certain problematic areas, such as hatched regions, still cannot be reliably detected. To address this, synthetically generated plans will be created and automatically segmented. This approach aims to enhance the AI model's ability to recognize hatched components and other challenging regions better. Following these refinements to the system components, an evaluation must be conducted to obtain generalizable results. The current experiments serve only as preliminary indicators and cannot draw definitive conclusions.

References

1. Lobbezoo, A., Qian, Y., Kwon, H.-J.: Reinforcement learning for pick and place operations in robotics: a survey, vol. 10, no. 3 (2021)
2. Neef, C., Luipers, D., Bollenbacher, J., et al.: Towards intelligent pick and place assembly of individualized products using reinforcement learning. In: Human Systems Engineering and Design III, vol. 1269 (2021)
3. Niu, H., Ji, Z., Zhu, Z., et al.: 3D vision-guided pick-and-place using kuka LBR iiwa robot (2021)
4. Berscheid, L., Meisner, P., Kroger, T.: Self-supervised learning for precise pick-and-place without object model, vol. 5, no. 3 (2025)
5. Seliger, R., Micheler, M., Gül-Ficici, S., et al.: Accelerating manual pick-and-place operations with AR-projected cad plans and ai-assisted object recognition. In: EUROCAST 2024: 19th International Conference (2024)
6. Seliger, R., Gül-Ficici, S., Göhner, U.: From paper to pixels: a multi-modal approach to understand and digitize assembly drawings for automated systems. In: Database and Expert Systems Appl. - DEXA 2024 Workshops, vol. 2169 (2024)
7. Ghalamzan, A.M.E., Mavrakis, N., et al.: Task-relevant grasp selection: a joint solution to planning grasps and manipulative motion trajectories. In: International Conference on Intelligent Robots and Systems (2016)
8. Fontanals, J., Dang-Vu, B.-A., et al:, Integrated grasp and motion planning using independent contact regions. In: 2014 IEEE-RAS International Conference on Humanoid Robots (2025)
9. Wiedholz, A., Heider, M., Nordsieck, R., et al.: Cad-based grasp and motion planning for process automation in fused deposition modelling (2021)
10. Ahmad, S., Samarawickrama, K., Rahtu, E., et al.: Automatic dataset generation from CAD for vision-based grasping. In: 2021 20th International Conference on Advanced Robotics (ICAR) (2021)
11. Labbé, Y., Manuelli, L., Mousavian, A., et al.: MegaPose: 6D pose estimation of novel objects via render & compare (2022)
12. Bauza, M., Bronars, A., Hou, Y., et al.: SimPLE, a visuotactile method learned in simulation to precisely pick, localize, regrasp, and place objects, vol. 9 (2025)
13. Wei, H., Pan, S., Ma, G., et al.: Vision-guided hand–eye coordination for robotic grasping and its application in tangram puzzles, vol. 2, no. 2 (2025)
14. Markaki, S., Panagiotakis, C.: Jigsaw puzzle solving techniques and applications: a survey, vol. 39, no. 10 (2023)
15. Yamada, F.M., Batagelo, H. C., Gois, J. P., et al.: TANGAN: solving tangram puzzles using generative adversarial network, vol. 55, no. 6 (2025)
16. Hugging Face: Idefics2: Next-generation open-weight multimodal models (2024). https://huggingface.co/blog/idefics2
17. Hu, H., Yang, J., Gao, F., et al.: mplug-docowl 1.5: Unified structure learning for ocr-free document understanding (2024)
18. JaidedAI: Easyocr (2020). https://github.com/JaidedAI/EasyOCR

Towards Real-Time Tool Wear Detection on Edge Devices: A Lightweight Dimensionality Reduction Approach for Spindle Integrated Cutting Force Sensor Data

Sebastian Unsin[✉], Benedikt Müller, Thomas Jäkel, and Frank Schirmeier

Institute for Data-Optimized Manufacturing (IDF), University of Applied Sciences Kempten, Kempten, Germany
`sebastian.unsin@hs-kempten.de`

Abstract. We present an efficient algorithm for monitoring milling tool wear using data from a spindle integrated Cutting Force Sensor. In a near-real-world milling experiment, we collected quality data of the manufactured part, tool wear, and cutting forces. A preprocessing pipeline transforms the data from time-domain to angle-domain and ensures its integrity. Using Singular Value Decomposition (SVD) for dimensionality reduction, we achieve a compact encoding of key sensor data and avoid manual intervention in the feature engineering. Notably, the SVD reconstruction error proves to be a reliable indicator for tool wear.

Keywords: Tool Wear Detection · Condition Monitoring · Dimensionality Reduction · Spindle Integrated Cutting Force Sensor · Milling Process

1 Introduction

Manufacturing industries continually strive for efficiency and precision, making tool wear detection essential for maintaining product quality and minimizing production downtime. The detection of tool wear traditionally relies on direct measurement techniques at the tool, which often prove impractical, as they interrupt the production process. A common strategy for ensuring the machining quality is changing the tools very early, regardless of the actual tool condition. This strategy is costly and unsustainable. Ongoing research in Predictive Maintenance and Tool Condition Monitoring is addressing the need for reliable and real-world capable tool wear detection.

Physically informed models combine process data with simulated cutting forces. [1] introduced a wear indicator based on cutting models, which was applicable to different cutting parameters and materials. However, it relies on a preprocessing of the G-code, removed geometry and a reference run with measured cutting forces, which makes it challenging for real-world applications. A comparative review on physically informed, data-driven and hybrid models was done by [2].

Artificial Intelligence driven approaches recently gathered a lot of interest and are covered by [3]. [4] compared machine learning techniques on acceleration sensor data.

A large survey on AI-methods for condition monitoring in milling can be found in [5]. AI-based systems are criticized for their need of large amounts of data and/or hardware resources to work; thus, the potential shown in scientific experiments often cannot be transferred to real world settings [6].

Advances in sensor technology have introduced new possibilities. Sensory tool holders capture bending moments, torque, and axial forces close to the tool, providing direct insights into tool defects through a rotating coordinate system. However, data analysis is complex due to transmission dropouts at high speeds, necessitating careful preprocessing. [7] investigated bending moments aggregated over time with a different process but noted increased moments during tooth breakage. [8] analyzed data mainly in the frequency domain using a self-developed sensory tool holder. Recent work by [9] applied dimensionality reduction using principal component analysis to detect cutting activity versus idle state, training a regression model on tool wear with manually specified features.

Dimensionality Reduction techniques [10, 11] are well-established in mathematics and machine learning. Popular methods include linear algebra-based techniques like Principal Component Analysis (PCA) and Singular Value Decomposition (SVD), as well as non-linear methods like t-SNE [12], UMAP [13] and Autoencoders [14].

This work continues the idea that dimensionality reduction is essential in a robust machine learning pipeline. However, it follows three main design principles to differentiate from previous studies:

1. We focus on bending moment polar data from spindle integrated cutting force sensors, considering it as most indicative for tool condition due to its rotating coordinate system.
2. During preprocessing, we minimize information loss and maximize the applicability to new and unseen operational scenarios by capturing the detailed shapes in bending moment polar plots that standard aggregations and predefined features miss.
3. We avoid supervised learning techniques due to challenges in obtaining reliable and comparable ground truth for tool wear. Wear measurements vary among researchers, limiting the generalizability of supervised models. Instead, we use unsupervised models to detect changes in tool condition compared to a reference period, aiming for robustness and applicability across diverse real-world scenarios.

In summary, we present a method to leverage rotational sensor data for wear detection using dimensionality reduction. This technique is efficient and utilizes standard linear algebra tools, enabling implementations on edge devices, such as machine tools or smart sensors.

2 Methods and Data Acquisition

This study builds upon insights and experimental data from the KI-Span research project [15], a collaborative effort involving renowned manufacturers across the machining chain and researchers in engineering and data science. The project conducted multiple machining experiments under laboratory as well as production settings.

We focus on two series of milling processes performed on a DMU65 FD monoBLOCK (DMG) machine tool using X5CrNi18–10 steel blocks (Ø120 mm ×

150 mm). Designed to closely mimic real-world conditions, these experiments involved various steps including roughing and finishing rather than monotonous block machining (see Fig. 1).

Fig. 1. Milling contour. Each run of the milling program includes 3 subsequent steps (blue, green, red) and is repeated until tool breakage is detected. After every step 2, surface quality and precision are measured at 6 points. The sample workpiece was designed by DMG MORI. (Color figure online)

The process was carried out using 12 tools across 91 runs. Cutting parameters remained constant for tools 1–5, while they varied for tools 6–12 to facilitate better generalization across different process conditions (listed in Appendix Table 2). A detailed description of the experimental setup will be provided in a separate publication.

Process Data: Cutting Forces. Process data was collected using a spindle integrated cutting force sensor, the pro-micron spike® Inspindle [16]. Figure 2 illustrates the axes for measuring cutting forces and presents a bending moment polar plot, the key data source for this study. Tools were positioned in the holder following a standardized procedure to ensure uniform placement.

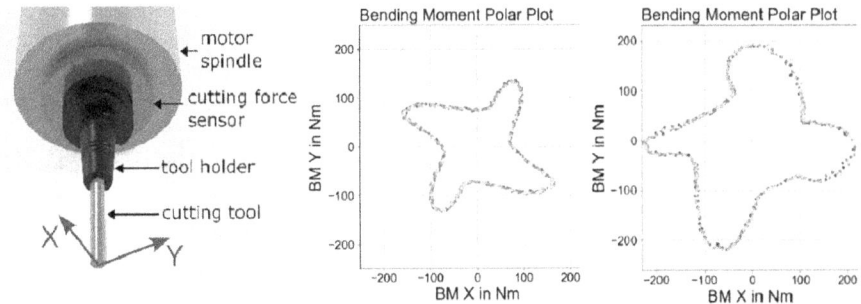

Fig. 2. Left: Schematic Diagram of a spindle integrated cutting force sensor with bending moment coordinates axes X and Y. Note that the X and Y axes of the bending moment rotate with the tool, thus the impact of a defect edge on the bending moment appears always in the same direction within the polar plot. Middle: Bending Moment Polar plot of an early life tool. The four extensions of the shape correspond to the four blades of the tool. Right: Bending Moment Polar Plot of a damaged tool. Tool damage is indicated by the irregular shape of the bending moment curve. The coloring indicates, that points are unordered within the timeframe of 0.2 s each.

Quality Data and Tool Wear Measurement. Within each run of the process, surface quality was assessed using Ra and Rz values measured at six different locations across the contour. Additionally, the inner and outer diameters and the contour width were measured to monitor dimensional accuracy. Tool wear was evaluated using images from a Keyence microscope and a standardized procedure to measure flank wear and notch wear.

An overview on process data, quality data and tool wear data are depicted in Fig. 9 of the results chapter. Remind that neither quality data nor tool wear measurements were used as target variables for the proposed tool wear detection approach. As we utilize unsupervised learning, these serve solely for validation purposes.

3 Data Preprocessing

The spindle integrated cutting force sensor outputs time series data for Axial Force, Torque, and Bending Moments in X and Y directions, resulting in approximately 70 million data points, including idle states.

In sensor signal processing, it is common to analyze timeframes within a window size that is sufficiently large to capture significant information but small enough to ensure consistent conditions. For milling, cutting conditions such as rotational speed and material removal rate should remain constant within these timeframes. We used a 0.2 s timeframe in our data, corresponding to about 500 data points and 11 rotations.

Preprocessing real-world data from the spindle integrated cutting force sensor involves addressing three major challenges:

1. **Retrieving the Manifold:** Bending moment polar data is a 2D point cloud which forms a non-linear 1D subspace, when cutting conditions remain constant within the timeframe (Fig. 2). As this manifold is unordered in time, we need a parametrization for efficient description.
2. **Non-stationarity**: Within the cutting program, the milling engagement conditions can vary greatly over time due to changing cutting conditions. Polar data from different timestamps distinguishes significantly in shape as well as in scale, which is not captured by simple aggregation features.
3. **Sensor dropouts and time anomalies:** High rotational speeds in data transmission can cause brief sensor data dropouts, leading to incomplete time index data and complicating the use of frequency-based features and algorithms.

Our solution to these challenges involves transforming the data from the time-domain to the angle-domain using an angle-related parametrization of the bending moments. This approach is robust to dropouts and retains maximum information.

Smoothed Bending Moment Angles. For each time (t), we denote the bending moment angle of a point $(x(t), y(t))$ in the polar plot relative to the origin as

$$\beta(t) = \tan^{-1} \frac{y(t)}{x(t)} \tag{1}$$

The angular velocity of bending moments is directly related to the spindle's rotation speed ω, meaning one mechanical rotation of the spindle equates to one rotation in bending moment coordinates. However, the actual rotation angle $\theta(t)$ of the machine tool cannot be tracked, because the spindle integrated cutting force sensor's rotation coordinates lack a fixed reference in workpiece coordinates. As an alternative, we define the smoothed bending moment angle $\alpha(t)$ as approximation of $\beta(t)$. If constant rotation velocity can be assumed during a timeframe, $\alpha(t)$ is a linear approximation of the unwrapped bending moment angle. Otherwise, smoothing splines can be used instead of linear regression, which was however not necessary for this work. Note that angles should be unwrapped before approximation and wrapped again afterwards to avoid disturbance by phase jumps. For each time window k, we define the approximated rotation angle α:

$$\alpha(t) = \alpha_0 + \omega_k t \qquad (2)$$

Observe that $\alpha(t)$ is an antiderivative to the true rotation velocity ω_k, as is as the true rotation angle $\theta(t)$. However, both differ in an unknown constant. The intercept α_0 of the linear approximation is chosen to fit $\beta(t)$ and keep its rotary coordinates smoothed in $\alpha(t)$.

Depending on tool geometry, the bending moment angle $\beta(t)$ may slightly lead or lag behind the smoothed bending moment angle $\alpha(t)$, as illustrated in Fig. 3. For similar reasons, $\beta(t)$ can locally decrease or remain constant while $\alpha(t)$ is constantly increasing. This makes $\beta(t)$ an invalid choice for the global parametrization of the bending moment manifold, as B(β) may be non-unique (Fig. 3 Top Right). In contrast $\alpha(t)$ allows a functional dependence to bending moments, as being closely related to the true rotation angle $\theta(t)$ (see Fig. 4 in next subsection).

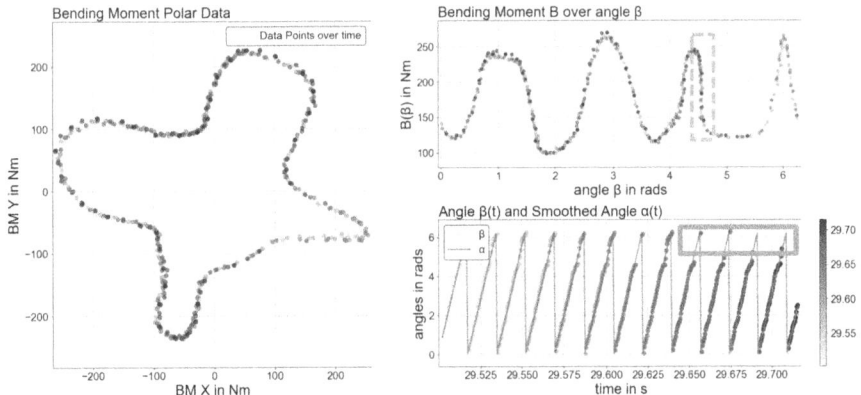

Fig. 3. Left: Bending moment polar plot in which the bending moments form a 1D manifold. The color progression from bright to dark indicates that the manifold is not ordered in the time. Right Top: Bending Moment B over β. In the dashed highlighted rectangle B(β) is not well-defined, e.g. for ß ≈ 4.5, the value of B is non-unique. Right Bottom: Corresponding scatter plot of $\beta(t)$ over time with approximated smoothed angle $\alpha(t)$ (red line). The phase jump in α indicates approximately 11 rotations within the timeframe. The blue rectangle highlights missing data in the last four rotations of the timeframe. (Color figure online)

The sensor used in our experiment occasionally recorded 1–3 samples with corrupted timestamps after a sensor dropout, leading to discontinuities in cutting force variables.

At this stage, these data points with corrupted timestamps can be filtered out, as their angle offsets $|\alpha - \beta|$ usually exceed 0.5, while typical offsets fall within [−0.5, 0.5]. We want to add that current versions of the sensor might not need this correction step.

Combining Multiple Rotations in Angle-Domain. For a bending moment datapoint (x,y), we define the absolute bending moment B via its Euclidean norm:

$$B(x(t), y(t)) := \sqrt{x(t)^2 + y(t)^2} \tag{3}$$

Having the smoothed angle α for each data point, we can consider the absolute bending moments as a function of α (see Fig. 4 for an example). This approach has the following advantages:

1. **Noise Reduction**: The radial basis function approximation (RBF) reduces sensor noise (which has the magnitude of approximately 1 Nm in most examples).
2. **Gap Filling**: Missing points, like those highlighted in Fig. 3, are augmented with data from other rotations, creating a gap-free representation.
3. **Process Consistency**: The quality of the fit indicates process consistency over rotations. Timeframes where the tool is moving in or out of the material result in bad approximation fits and can be filtered out. We retain timeframes where the normalized residual root mean square error is less than 0.06.

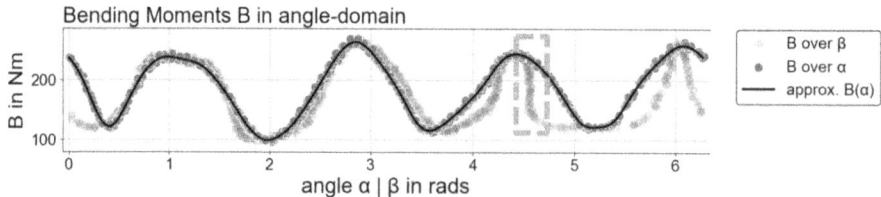

Fig. 4. Absolute bending moment B in the angle domain, with red points mapped over the smoothed angle α. For comparison, B is shown additionally over the original bending moment angle (β). In the highlighted region around ß ≈ 4.5, the original data does not allow a functional dependency over (β), resulting in non-unique mapping B(β). Note that the smoothed angle α does not suffer from this issue.

The preprocessing is crucial as a foundation for further analysis. We summarize our preprocessing pipeline in five steps:

1. Partition Data in timeframes (here: 0.2 s)
2. Compute smoothed angle α as approximation of bending moment angle β (here: linear regression on unwrapped angles)
3. Remove data points with large angle offset (here: $|\alpha - \beta| > 0.5$), in case these represent more than 50% skip the whole timeframe
4. Sort data by smoothed angle α and approximate absolute bending moment B(α) (here: RBF-interpolation)
5. Skip the whole timeframe, if approximation error is too large (here: normalized residual root mean square error < 0.06)

At the end, around 40% cleaned and consistent timeframes remained. A similar transformation method is described as Time Synchronous Averaging (TSA) [17]. Further details and additional transformation methods can be found in [18].

4 Dimensionality Reduction

The described preprocessing steps converted our data set to angle-domain data in matrix form (polar matrix P) with shape (53030, 360), where 53030 represents the number of timeframes and 360 is the angular resolution (1°). Using randomly chosen rows (time frames), Fig. 5 exemplifies that the overall patterns are similar, but contain subtle distinctions, which we argue can be associated to varying tool wear conditions.

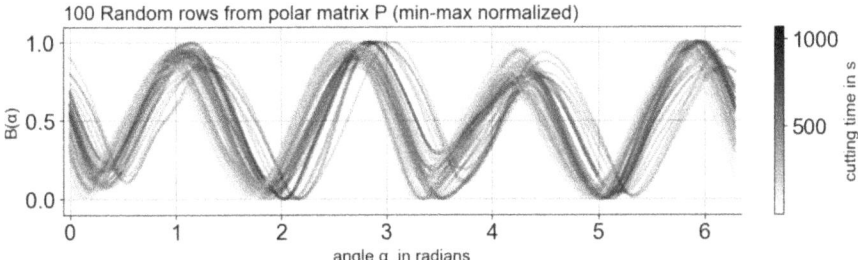

Fig. 5. Comparison of 100 randomly chosen rows of data matrix P. The angle-domain data shows similarities, with samples having higher cutting times (dark colors) slightly shifted to the right.

Dimensionality reduction techniques can effectively condense the information into fewer variables (columns). Our intention is not to compress data size but rather to find a compact representation for wear detection algorithms. Due to the simplicity of our data patterns, we focused on linear dimensionality reduction methods for their robust theoretical basis, interpretability, and efficient implementations suitable for edge devices. We preferred SVD over PCA, as it offers similar performance and slightly easier interpretation.

Singular Value Decomposition (SVD). Singular Value Decomposition (SVD) is a fundamental matrix factorization technique in linear algebra [19]. It decomposes any m × n matrix A into three matrices:

$$A = U\Sigma V^T \tag{4}$$

where U is an m × m orthogonal matrix whose columns are the left singular vectors of A, V is an n × n orthogonal matrix whose columns are the right singular vectors of A and Σ is an m × n diagonal matrix containing the singular values (σ_i), which are non-negative real numbers.

For our data, (A = P), the right singular vectors have a dimension of 360 and encode functions in the angle-domain. Typically, only a few singular values are large and significant, while many others are small and can be considered as negligible. Efficient algorithms avoid doing a full singular value decomposition and iteratively compute only d singular values (*truncated SVD*).

$$A = U\Sigma V^T = \sum_{i=0}^{r} \sigma_i u_i v_i^T \approx \sum_{i=0}^{d} \sigma_i u_i v_i^T \qquad (5)$$

Hereby r denotes the rank of A and d is the reduced dimension, which is chosen typically much smaller than r. The singular vectors corresponding to the largest singular values contain the directions in which the data varies the most. In our case, these singular vectors describe typical patterns of bending moments in the angle-domain, as visualized in Fig. 6 left. Three components were sufficient to capture approximately 99% of the variance in the data. Once the decomposition is done, the *reconstruction error R* of a k-dimensional vector p (bending moment on angle-domain) is the residual root mean square error (RMSE) of the transformed and again inverse-transformed SVD:

$$R(p) = \left\| \frac{1}{k}(p - \widehat{p}) \right\|_2 = \left\| \frac{1}{k}\left(p - \left(VS^{-1}U^T \cdot \left(USV^T p\right)\right)\right) \right\|_2 \qquad (6)$$

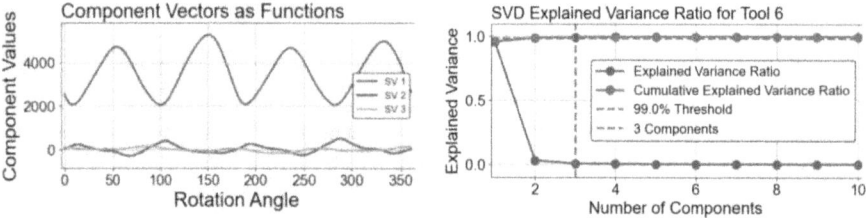

Fig. 6. Left: SVD components (singular vectors) plotted over the smoothed angle. SV1 captures most of the variance and shows a curve similar to the samples in Fig. 5. SV2 and SV3 encode the most common deviations from the shape of SV1, allowing linear combinations of SV1, SV2, and SV3 to closely resemble most bending moment polar data. Right: Explained variance for the first 10 SVD components. Using just 3 components captures over 99% of the total variance.

5 Tool Wear Detection

Exploratory analysis showed that tool wear causes irregular bending moment polar data. These irregularities are harder to express in a dimensionality-reduced setting, especially when the dimensionality reduction has been optimized on a data set in which most observations are regular. Thus, we derived the following tool wear detection approach:

1. Apply SVD to the angle-domain data from a reference timespan (calibration phase) where the tool condition is known to be regular.
2. Compute the truncated SVD representation for each subsequent timespan using components from the calibration phase.
3. Re-transform the reduced data back and compare it to the uncompressed angle-domain data.

4. Compute the reconstruction error R and compare it to the average reconstruction error from the calibration phase.
5. Indicate a tool wear warning if the reconstruction error is significantly larger compared to the calibration phase.

The calibration phase should cover all process steps at least once and be sufficiently short to avoid tool wear. All operations in this pipeline are linear algebra operations, which can be efficiently processed. On a regular CPU, transforming and re-transforming data for a complete tool takes less than 2 ms.

Tool wear assessment is independent of the tool's full operational history; instead, each timeframe is evaluated only against the calibration phase of the same tool. Despite minor fluctuations, the unsmoothed reconstruction error trajectory remains robust, even under complex process conditions. Figure 7 illustrates the algorithm for one tool.

Comparing reconstructed and real polar plots provides further insight (Fig. 8). The SVD components fit the data well during the calibration phase but fail to capture the polar shape after extended cutting time. By the lifetime's end, the true polar plot has rotated nearly 30 degrees, resulting in significant reconstruction mismatch. A similar effect was reported by [20] both on simulated and measured cutting forces.

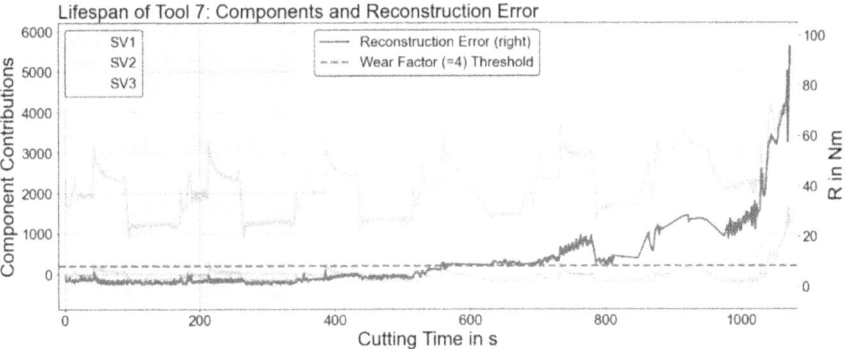

Fig. 7. Component contribution and reconstruction error over the lifetime of a tool. The calibration phase is marked with a grey background. Light-colored lines show singular value components, with SV1 (blue) being the most significant. At the end of the lifetime, SV2 and SV3 also increase. The reconstruction error, linked to the right y-axis, ranges around 2–3 Nm during the calibration phase. During the post-calibration phase, it rises, indicating a wear warning around x = 550s. (Color figure online)

6 Validation and Results

During the experiment, each tool's lifetime was monitored, with measurements of workpiece quality and tool wear evaluation between process steps, as described in Sect. 2. Both quality and tool wear data served as unbiased test data, not used for training the model or calculating reconstruction error. The SVD was computed on each tools calibration phase

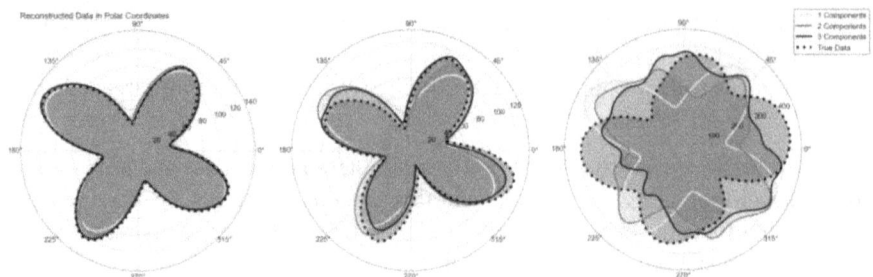

Fig. 8. Reconstructed vs. true bending moment polar data at the beginning and end of tool 7's life (SVD1: yellow, SVD1+2: blue, SVD1+2+3: purple, True Polar: dotted). Left: The true data fits well with 3 SVD components initially. Middle: Slight mismatch in the middle of tool life. Right: At the end of the tool's life, the true bending moment polar data differs in shape and orientation. The component shapes remain unchanged, thus no linear combination of SVD components fits, leading to high reconstruction mismatch. (Color figure online)

separately and the reconstruction error by forward and backward transformation of later timeframes. It is subject to future work to study the possibility to transfer a trained tool wear indicator to new and unseen tools.

Given the realistic cutting parameters and high-quality tools, the lifespans were long, and only 12 tools were worn. Evaluation was conducted both qualitatively, using graphs (e.g. Fig. 9) and against predefined validation criteria for quality data. Table 1 outlines the tolerances on quality criteria and tool wear and the indicator R.

Table 1. Tolerances for evaluating tool change recommendations. The last column indicates the tolerance for the wear indicator, based on the average reconstruction error during the calibration phase increase factor.

	Flank wear	Notch Wear	Rz	Contour width	Reconstruction R
Tolerance	300 μm	500 μm	8 μm	20 μm	increase factor: 3

Each tool's critical wear indication time was compared to the point when any validation tolerance criterion was exceeded. For 9 of the 12 tools, the wear indicator suggested tool changes shortly or reasonably short before quality tolerance exceedance. For tools 10 and 12, the wear indication was very early. For tool 10, this could be due to a process anomaly. For tool 12, the indicator warned too late, but qualitative data suggests the contour width tolerance exceedance might be a measurement artifact due to temperature increase, as it was only a temporary increase. Details are listed in Appendix Table 3.

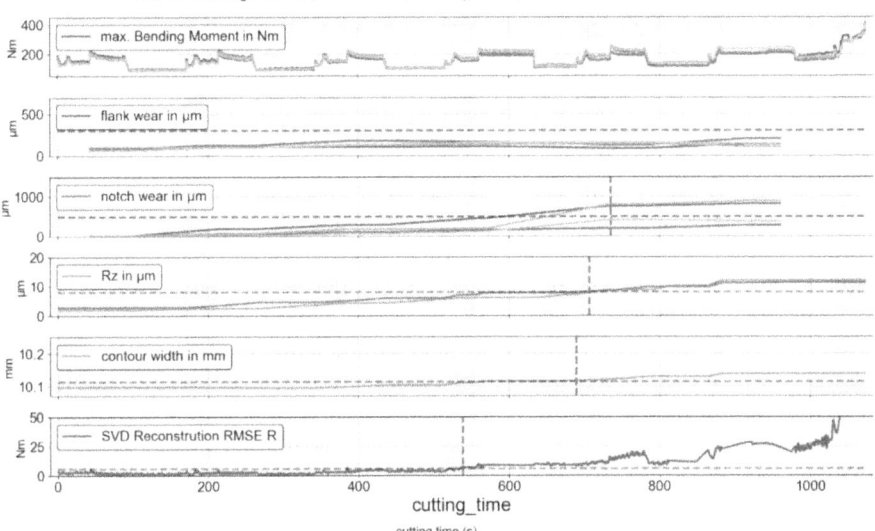

Fig. 9. Overview of measured data for tool 7 across its entire lifespan in 7 runs. Different process steps are indicated by background colors (red, yellow, green). The first row shows the maximal bending moment per 0.2 s timeframe, with colors indicating contributions from the 4 different edges visible in the polar plot. The second and third rows depict flank wear and notch wear for each of the 4 edges, available only after each run. The fourth and fifth rows display quality data for Rz surface values and contour width precision. The last row shows the SVD reconstruction error-based wear indicator R, with tolerance level marked as dashed line. The wear indicator reaches its tolerance shortly before quality criteria exceedance, which is desired. (Color figure online)

7 Discussion

Key Findings

This study demonstrates the successful use of data from spindle integrated cutting force sensors to monitor milling processes. We provide a robust preprocessing pipeline leading to an elegant angle-domain representation. This angle-domain data proves to be well-suited for further analysis, showing resilience against missing data, noise, and variations in rotational speed. The timeframes in angle-domain act as an excellent basis for dimensionality reduction using SVD or PCA. We showed that only a few dimensions are sufficient to describe the data. Tool wear detection algorithms can efficiently work on these reduced dimensions, capturing most relevant information. As a promising result, we demonstrated that the SVD reconstruction error, compared to a reference timespan, is a computationally lightweight but reliable tool wear indicator.

Further Insights and Future Perspectives

Stacking Equal and Individual Tool Components. In this study each tool was treated separately, as previous work indicated substantial exemplar variance even among high-quality tools. However, a calibration-free tool wear indicator would be highly desirable. Combining SVD models from previously analyzed identical tools and adding specific components during the initial seconds of a newly beginning process (e.g. adjusting for variations in the tool rotation relative to toolholder) could enhance wear prediction without calibration.

Unsupervised Learning on SVD Component Values. While dimensionality reduction lays the groundwork, more sophisticated machine learning algorithms could refine the analysis. Evaluating the distributions of all singular value components could lead to a detailed model distinguishing normal and worn tool conditions. We propose using unsupervised learning during a calibration phase to learn the distribution of regular tool conditions, comparing real-time data to this model trained during the calibration phase. Initial work employing Gaussian Mixture Models show promising results.

Including Process Context Information. In real-world processes, conditions such as material removal rates vary greatly. Integrating context information into machine learning models could significantly improve data analysis. For our unsupervised learning approach, we include the relative time in process as variables. More precise context information, like synchronized spindle coordinates, would further enhance accuracy. Alternatively, similar cutting parameters can be derived from a removal simulation like done by [21].

Involvement of Angular Offset During Preprocessing and Dimensionality Reduction. Transforming bending moment angles to linearized smoothed angles sometimes masks abnormalities, which were present in the original data. To address this, encoding the angle-offset as a phase of a complex-valued bending moment preserves this information in an elegant representation. Although initial tests showed poorer indication from reconstruction errors, the angle-offset might offer valuable additional data for more sophisticated algorithms.

Online In-process SVD to Avoid the Storage of Data on Edge-Devices. If edge devices cannot store complete calibration phase data, incremental SVD algorithms offer a solution through online learning [22]. Initial values from identical tool data can serve as a warm start for this process.

Analysis of the Bending Moment Angle Rotation. In exploratory analysis, increased tool wear – and particularly tool failure – causes the bending moment polar data to rotate anti-clockwise. This phenomenon supports the effectiveness of the reconstruction error-based indicator and could be valuable for understanding wear processes as well as serving as another lightweight and interpretable wear indicator.

With this study we want to contribute to the effective data use in tool condition monitoring, particularly when using a spindle integrated cutting force sensor. We emphasize the advantage of angle-domain data and dimensionality reduction for efficient analysis and aim to enable masurable cost reductions and efficiency gains in the manufacturing industry.

Acknowledgments. The project "AI-supported optimization of tool life and component quality on machine tools in machining production (KI-Span)" (original: "KI-gestützte Optimierung der Werkzeugstandzeit und der Qualität der Bauteile an Werkzeugmaschinen in der spanenden Fertigung (KI-Span)"), running from August 2021 to December 2024, was funded under the Bavarian Collaborative Research Program (BayV-FP; funding line: Digitization) by the Bavarian Ministry of Economic Affairs, Regional Development and Energy (StMWi).

Disclosure of Interests. The authors have no competing interests to declare that are relevant to the content of this article.

Appendix

Table 2. Overview of Cutting Parameters for each tool of the experiment. The tools were produced by Ceratizit and for tool 11 and 12 the tool choice was intentionally unconventional for the material. The coloring indicates, if process experts considered the cutting parameter being "tool friendly" (green), tool challenging (yellow) or untypical (red).

Tool Number	vc	n	D	f	vf	ap	ae	Article Number
	m/min	1/min	mm	mm/U	mm/min	mm	mm	Ceratizit Tool #
1-5	115	3050	12	0,4	1220	8	3	5407112200
6-7	140	3714	12	0,4	1486	8	3	5407112200
8-9	130	3448	12	0,5	1724	8	3	5407112200
10	130	3448	12	0,24	828	8	4	5407112200
11	130	3448	12	0,24	828	8	4	5250412000
12	110	2918	12	0,32	934	8	2	5250412000

Table 3. Comparison of Validation Tolerance Exceed and Indicator Exceed for each tool (rows) w.r.t. cutting time in seconds. Second row shows when quality data indicates tool wear beyond predefined tolerance. Third column shows, when the tool wear was indicated by exceeded reconstruction error tolerance. The difference column shows prewarning time (indicator to quality issue) in seconds. For some tools a comment is added for explanation.

Tool Number	Quality Tolerance exceeded	Indicator Tolerance exceeded	Difference (seconds)	Comment
1	1174	784	390	
2	1433	1120	313	
3	1192	1033	159	
4	1043	1492	-449	measurement issue (Temp. Change)
5				
6	387	386	1	
7	689	550	139	
8	496	468	28	

(*continued*)

Table 3. (*continued*)

Tool Number	Quality Tolerance exceeded	Indicator Tolerance exceeded	Difference (seconds)	Comment
9	314	338	-24	quality data interpolated between runs
10	901	175	726	possibly process anomaly
11	636	437	199	
12	2437	1336	1101	very long running tool

References

1. Nouri, M., Fussell, B.K., Ziniti, B.L., Linder, E.: Real-time tool wear monitoring in milling using a cutting condition independent method. Int. J. Mach. Tools Manuf. **89**, 1–13 (2015). https://doi.org/10.1016/J.IJMACHTOOLS.2014.10.011
2. Zhang, H., Jiang, S., Gao, D., et al.: A review of physics-based, data-driven, and hybrid models for tool wear monitoring. Machines **12** (2024). https://doi.org/10.3390/MACHINES12120833
3. Cheng, Y., Lu, M., Gai, X.: Research on multi-signal milling tool wear prediction method based on GAF-ResNext. Robot. Comput. Integr. Manuf. **85** (2024). https://doi.org/10.1016/J.RCIM.2023.102634
4. Moore, J., Stammers, J., Dominguez-Caballero, J.: The application of machine learning to sensor signals for machine tool and process health assessment. Proc. Inst. Mech. Eng. B J. Eng. Manuf. **235**, 1543–1557 (2021). https://doi.org/10.1177/0954405420960892
5. Pimenov, D.Y., Bustillo, A., Wojciechowski, S.: Artificial intelligence systems for tool condition monitoring in machining: analysis and critical review. J. Intell. Manuf. **34**, 2079–2121 (2023). https://doi.org/10.1007/S10845-022-01923-2
6. Liu, D., Liu, Z., Wang, B.: 2024 Leveraging artificial intelligence for real-time indirect tool condition monitoring: From theoretical and technological progress to industrial applications Int. J. Mach. Tools Manuf 202 https://doi.org/10.1016/J.IJMACHTOOLS.2024.104209
7. Veiga, F., Val, A.G., Del, Suárez, A., Alonso, U.: Analysis of the machining process of titanium Ti6Al-4V parts manufactured by wire arc additive manufacturing (WAAM). Materials **13**, 766 (2020). https://doi.org/10.3390/MA13030766
8. Gent, S., Gert, O., Schorghofer, P., et al.: Maintenance interval monitoring and cutting edge breakout detection using an instrumented tool. In: IEEE International Conference on Emerging Technologies and Factory Automation, ETFA September 2022 (2022). https://doi.org/10.1109/ETFA52439.2022.9921282
9. Pichler, K., Huemer, M., Kaineder, G., et al.: Wear detection for a cutting tool based on feature extraction and multivariate regression. In: Proceedings - 2024 IEEE International Conference on Information Reuse and Integration for Data Science, IRI 2024, 90–95 (2024). https://doi.org/10.1109/IRI62200.2024.00030
10. Van Der Maaten, L.J.P., Postma, E.O., Van Den Herik, H.J.: Dimensionality Reduction: A Comparative Review

11. Anowar, F., Sadaoui, S., Selim, B.: Conceptual and empirical comparison of dimensionality reduction algorithms (PCA, KPCA, LDA, MDS, SVD, LLE, ISOMAP, LE, ICA, t-SNE). Comput. Sci. Rev. **40**, 100378 (2021). https://doi.org/10.1016/J.COSREV.2021.100378
12. Maaten, L., Van Der Hinton, G.: Visualizing data using t-SNE J. Mach. Learn. Res. **9**, 2579–2605 (2008).
13. McInnes, L., Healy, J., Melville, J.: UMAP: Uniform Manifold Approximation and Projection for Dimension Reduction (2018)
14. Hinton, G.E., Salakhutdinov R.R.: Reducing the dimensionality of data with neural networks. Science 1979 **313**, 504–507 (2006). https://doi.org/10.1126/SCIENCE.1127647/SUPPL_FILE/HINTON.SOM.PDF
15. Unsin, S., Dorer, C., Müller, B., et al.: Schritt für Schritt zur Zerspanungs-KI. maschinenbau **4**, 10–15 (2024). https://doi.org/10.1007/S44029-024-1209-1
16. spike® – sensory tool holder made in Germany – pro-micron. https://www.pro-micron.de/en/spike-the-smart-tool-control-system-made-in-germany/. Accessed 30 Mar 2025
17. Bechhoefer, E., Kingsley, M.: A review of time synchronous average algorithms. In: Annual Conference of the PHM Society 1 (2009)
18. Du, N.T., Dien, N.P.: Advanced signal decomposition methods for vibration diagnosis of rotating machines: a case study at variable speed. In: Tien Khiem, N., Van Lien, T., Xuan Hung, N. (eds.) Modern Mechanics and Applications. LNME, pp. 393–400. Springer, Cham (2022). https://doi.org/10.1007/978-981-16-3239-6_30
19. Golub, G.H., Reinsch, C.: Singular value decomposition and least squares solutions Handb. Automat. Comput. 134–151 (1971). https://doi.org/10.1007/978-3-642-86940-2_10
20. Tseng, H.C., Tsai, M.S., Yeh, B.C., Li, K.M.: Analysis of tool wear by using a cutting bending moment model for milling processes. Int. J. Precis. Eng. Manuf. **23**, 943–955 (2022). https://doi.org/10.1007/S12541-022-00680-9/TABLES/2
21. Götz, M., Rost, M., Wilkner, D., Schirmeier, F.: Unsupervised segmentation of CNC milling sensor data into comparable cutting conditions. In: Strauss, C., Amagasa, T., Manco, G., Kotsis, G., Tjoa, A.M., Khalil, I. (eds.) DEXA 2024. LNCS (LLNAI and LNB), vol. 14911, pp. 149–155, Springer, Cham (2024). https://doi.org/10.1007/978-3-031-68312-1_12
22. Brand, M.: Incremental singular value decomposition of uncertain data with missing values. In: Heyden, A., Sparr, G., Nielsen, M., Johansen, P. (eds.) ECCV 2002. LNCS (LNAI and LNB), vol. 2350, pp. 707–720. Springer, Cham (2002). https://doi.org/10.1007/3-540-47969-4_47

Energy Optimized Piecewise Polynomial Approximation Utilizing Modern Machine Learning Optimizers

Hannes Waclawek[1,2] and Stefan Huber[1]

[1] Josef Ressel Centre for Intelligent and Secure Industrial Automation, Salzburg University of Applied Sciences, Salzburg, Austria
{hannes.waclawek,stefan.huber}@fh-salzburg.ac.at
[2] Paris Lodron University Salzburg, Salzburg, Austria

Abstract. This work explores an extension of Machine Learning (ML)–optimized Piecewise Polynomial (PP) approximation by incorporating energy optimization as an additional objective. Traditional closed-form solutions enable continuity and approximation targets but lack flexibility in accommodating complex optimization goals. By leveraging modern gradient descent optimizers within TensorFlow, we introduce a framework that minimizes elastic strain energy in cam profiles, leading to smoother motion. Experimental results confirm the effectiveness of this approach, demonstrating its potential to Pareto-efficiently trade approximation quality against energy consumption.

Keywords: Piecewise Polynomials · Gradient Descent · Approximation · TensorFlow · Electronic Cams · Energy Optimization

1 Introduction

1.1 Motivation and Problem Statement

PPs are commonly utilized in various scientific and engineering disciplines. One key area of interest is trajectory planning for machines in mechatronics, computing time-dependent position, velocity, or acceleration profiles that provide setpoints for controllers of industrial machines. In this context, electronic cams define the repetitive motion of servo drives, commonly defined as input point clouds and approximated by PPs. Traditionally, the approximation process involves solving a system of equations in closed form, incorporating domain-specific constraints like continuity, cyclicity, or periodicity. While computationally efficient, closed-form solutions offer limited flexibility regarding polynomial degrees, polynomial bases, or the integration of additional constraints.

As optimization goals grow more complex, numerical methods such as gradient descent become beneficial. Since gradient-based optimization lies at the heart of training (deep) neural networks, modern ML frameworks like TensorFlow or

PyTorch come with a range of recent optimizers. In previous work [1,5], we utilize these optimizers directly to fit $\mathcal{C}^k$-continuous PPs for the use in the electronic cam domain.

In this work we extend the problem setting to a constrained multi-objective optimization problem by adding an energy term as a domain-specific loss term and investigate (i) convergence and (ii) the pareto front under this extension. More precisely, given input points $x_1 \leq \cdots \leq x_n \in \mathbb{R}$ and respective target values $y_1, \ldots, y_n \in \mathbb{R}$, we look for a PP function f that minimizes both the approximation loss ℓ_2 and energy loss $\ell_E = \int_I f''(x)^2 \mathrm{d}x$ over some domain I that captures the total curvature, under the constraint that $f \in \mathcal{C}^k$, where $\mathcal{C}^k$ denotes the set of k-times continuously differentiable functions $\mathbb{R} \to \mathbb{R}$. We form a loss ℓ_{CK} that captures the "amount of violation" of $\mathcal{C}^k$-continuity, which will lead to the optimization constraint $\ell_{\mathrm{CK}} = 0$, as it is of practical relevance to feed servo drives with PPs of continuity class $\mathcal{C}^k$ to prevent excessive forces, vibrations and wear on machine parts.

Note that ℓ_E is related to energy in multiple ways: In a mechanical setting, ℓ_E is the elastic strain energy of a rod bent as the function graph of f. When f is used as a cam and we approximately assume that the acceleration f'' is proportional to the motor coil current then ℓ_E is proportional to the copper losses in the motor coils.

1.2 Prior and Related Work

Related work in the cam approximation domain mostly either relies on closed-form solutions, like [2], or utilizes parametric functions for approximation of given input data, like B-Spline or NURBS curves, as in [3]. The first is limited with respect to complexity and number of optimization goals and the latter cannot directly be utilized in industrial servo drives utilizing non-parametric PP functions. Our current work represents a natural progression from the principles introduced in [1,5].

2 The Optimization Problem

Parameterized model. Let us consider n input points $x_1 \leq \cdots \leq x_n \in \mathbb{R}$ with respective target values $y_1, \ldots, y_n \in \mathbb{R}$. We split the input domain $I = [x_1, x_n]$ into m sub-intervals $I_i = [\xi_{i-1}, \xi_i]$, with $1 \leq i \leq m$, where $x_1 = \xi_0 \leq \cdots \leq \xi_m = x_n$. A PP function f of degree d is defined by m polynomial functions $p_1, \ldots, p_m$ of degree d as

$$p_i(x) = \sum_{j=0}^{d} \alpha_{i,j} x^j, \qquad (1)$$

where p_i covers the sub-domain I_i of I. As a parameterized model, f possesses $(d+1) \cdot m$ model parameters $\alpha_{i,j}$, which are trained via gradient descent optimizers according to the loss functions described below.

Loss Functions. We build on [1] for the approximation loss ℓ_2 and the $\mathcal{C}^k$-continuity loss ℓ_{CK}. The former is defined as

$$\ell_2 = \frac{1}{n}\sum_{i=1}^{n} |f(x_i) - y_i|^2 \tag{2}$$

and the latter introduces a term $\Delta_{i,j}$ that measures the discontinuity of the j-th derivative at ξ_i to then define

$$\ell_{\mathrm{CK}} = \frac{1}{m-1}\sum_{i=1}^{m-1}\sum_{j=0}^{k} \Delta_{i,j}^2 \quad \text{with} \quad \Delta_{i,j} = p_{i+1}^{(j)}(\xi_i) - p_i^{(j)}(\xi_i). \tag{3}$$

Recall that the ξ_i are the boundary points of the domains I_i of p_i. Note that achieving $\ell_{\mathrm{CK}} = 0$ means that f is $\mathcal{C}^k$-continuous and $\ell_2 = 0$ means zero approximation error.

We can account for cyclicity (e.g., of the cam profile) by replacing in Eq. (3) the two occurrences of $m-1$ by m and generalizing $\Delta_{i,j}$ to

$$\Delta_{i,j} = p_{1+(i \bmod m)}^{(j)}(\xi_{i \bmod m}) - p_i^{(j)}(\xi_i). \tag{4}$$

(We may exclude the case $j = 0$ when $i = m$ if we do not require periodicity, see [1] for details.)

Energy loss. In this work, we motivated the energy loss ℓ_E for f as

$$\ell_E = \int_I f''(x)^2 \, \mathrm{d}x = \sum_{i=1}^{m} \int_{I_i} p_i''(x)^2 \, \mathrm{d}x \tag{5}$$

in the introduction and further note that

$$\int_{I_i} p_i''(x)^2 \, \mathrm{d}x = \int_{I_i} \left(\sum_{j=2}^{d} \alpha_{i,j} \cdot j(j-1) x^{j-2}\right)^2 \mathrm{d}x. \tag{6}$$

Solving the definitive integral in Eq. (6) and plugging it into Eq. (5) gives us the energy loss in closed form as a function of the trainable model parameters:

$$\ell_E = \sum_{i=1}^{m}\sum_{j=2}^{d}\sum_{k=2}^{d} \alpha_{i,j}\alpha_{i,k} \frac{jk(j-1)(k-1)}{j+k-3} \cdot \left(\xi_i^{j+k-3} - \xi_{i-1}^{j+k-3}\right). \tag{7}$$

Multi-objective Optimization. Various applications in engineering require $\mathcal{C}^k$ continuity up to some k, i.e., continuous velocities or accelerations of cam profiles. Hence, we treat $\ell_{\mathrm{CK}} = 0$ as a constraint in the 2-objective optimization problem

$$\begin{aligned}\min \quad & \beta \ell_2 + (1-\beta)\ell_E \\ \text{s.t.} \quad & \ell_{\mathrm{CK}} = 0\end{aligned} \tag{8}$$

over the trainable model parameters $\alpha_{i,j}$ and with $0 \leq \beta \leq 1$. We generalize [1] by turning the problem into an unconstrained 3-objective optimization problem

$$\min \quad \alpha \ell_{\text{CK}} + \beta \ell_2 + (1 - (\alpha + \beta))\ell_{\text{E}} \qquad (9)$$

with $\alpha, \beta \geq 0$ and $\alpha + \beta \leq 1$. Note that we can therefore apply the unconstrained optimizers shipped with ML frameworks like TensorFlow. The optimization result is then processed by the algorithm CKMIN from [5] to strictly establish $\ell_{\text{CK}} = 0$, which effectively adds correction polynomials to obtain $\mathcal{C}^k$-continuity.

3 Experimental Results and Discussion

The code and results discussed in this section along with further experiments and plots are provided in the public repository [4].

For our experiment, we consider a dataset with $n = 100$ input points $x_i \in [0, 1]$ and target values $y_i = \sin(4\pi x_i^2) + n_i$, where n_i is a normal noise with zero mean and standard deviation 0.1, produced by *numpy.random.normal*. In practice, such data could stem from measurements. In Fig. 1 the dataset is illustrated and we observe that solely optimizing for approximation loss leads to significant curvatures (left), while adding ℓ_{E} regularizes the result (right).

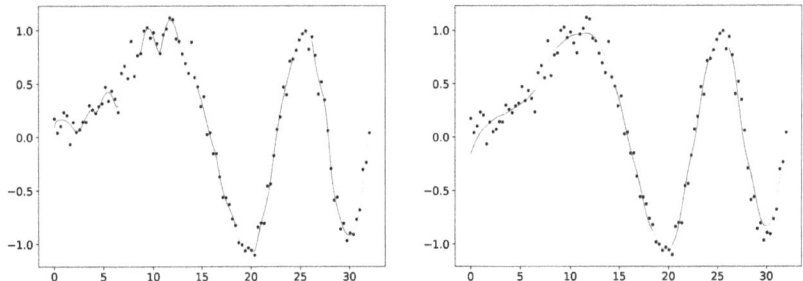

Fig. 1. Results for the first two rows of Table 1. Left: $\beta = 0.9$, right: $\beta = 0.45$.

We utilize the AMSGrad optimizer in TensorFlow to fit our PP model as explained in Sect. 2. Within our training loop, we run 1000 epochs with early stopping and patience of 100 epochs to fit a $\mathcal{C}^2$-continuous, periodic PP of degree 7 with 16 polynomial segments. Table 1 and Fig. 2 show the results after establishing $\ell_{\text{CK}} = 0$ using algorithm CKMIN from [5] for different α and β. Table 1 confirms that additionally optimizing ℓ_{E} works: Reducing β from 0.9 to 0.45, shown in the first two rows of Table 1, reduces ℓ_{E} by a factor of 4.40.

4 Conclusion

Our results demonstrate that an extension of the framework introduced in [5] and [1] towards energy optimization is feasible. Results converge without additional

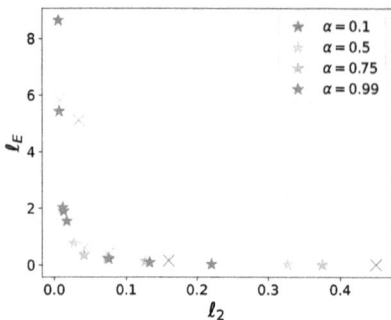

Fig. 2. ℓ_E over ℓ_2 for different α and $\beta \in (0, 1-\alpha]$. ⋆ is Pareto optimal.

Table 1. Results for different α, β.

Parameters	ℓ_2	ℓ_E
$\alpha = 0.10, \beta = 0.900$	0.005	8.682
$\alpha = 0.10, \beta = 0.450$	0.014	1.970
$\alpha = 0.50, \beta = 0.500$	0.007	5.468
$\alpha = 0.50, \beta = 0.250$	0.046	0.616
$\alpha = 0.75, \beta = 0.250$	0.007	5.852
$\alpha = 0.75, \beta = 0.125$	0.077	0.282
$\alpha = 0.99, \beta = 0.010$	0.043	4.977
$\alpha = 0.99, \beta = 0.005$	0.135	0.121

regularization measures and indeed improve in the case of input data that is unfavorable for sole approximation and continuity optimization. Optimizing ℓ_E further has positive side effects: Fig. 1 shows that oscillations in the curve's shape are reduced, which is favorable from a mechanical perspective.

While orthogonal bases, like Chebyshev polynomials, have shown much better performance on ℓ_2, preliminary tests indicated convergence is impaired when adding ℓ_E. We leave the investigation of mitigation measures to future work.

Acknowledgments. This work was supported by the Christian Doppler Research Association (JRC ISIA) and the European Interreg Österreich-Bayern project BA0100172 AI4GREEN. We want to thank the reviewers for their valuable feedback.

Disclosure of Interests. The authors have no competing interests to declare that are relevant to the content of this article.

References

1. Huber, S., Waclawek, H.: $mathcalC^k$-continuous spline approximation with tensorflow gradient descent optimizers. In: Moreno-Díaz, R., Pichler, F., Quesada-Arencibia, A. (eds.) EUROCAST 2022. LNCS, vol. 13789, pp. 577–584. Springer, Cham (2023). https://doi.org/10.1007/978-3-031-25312-6_68
2. Mermelstein, S., Acar, M.: Optimising cam motion using piecewise polynomials. Eng. Comput. **19**, 241–254 (2004). https://doi.org/10.1007/s00366-003-0264-0
3. Nguyen, T., Kurtenbach, S., Hüsing, M., Corves, B.: A general framework for motion design of the follower in cam mechanisms by using non-uniform rational B-spline. Mech. Mach. Theory **137**, 374–385 (2019)
4. Waclawek, H.: Experimental Results (2025). https://github.com/JRC-ISIA/paper-2025-energy-ml-optimized-pp. Accessed 09 Apr 2025
5. Waclawek, H., Huber, S.: Machine learning optimized orthogonal basis piecewise polynomial approximation. In: Festa, P., Ferone, D., Pastore, T., Pisacane, O. (eds.) LION 2024. LNCS, vol. 14990, pp. 427–441. Springer, Cham (2025). https://doi.org/10.1007/978-3-031-75623-8_33

Author Index

A
Adami, Cornelia 30

B
Belinskaya, Yulia 15
Bencik, Batuhan 30
Brucker, Mathias 30
Buhl, Cindy 73

C
Chirakal, Teena 47

E
Eigner, Oliver 15
Ekaputra, Fajar J. 3

F
Fischer, Lukas 30

G
Göhner, Ulrich 73, 79, 87
Gül-Ficici, Sebnem 87

H
Havas, Clemens 3
Heieck, Frieder 47
Hirlaender, Simon 67
Huber, Florian 47
Huber, Stefan 67, 110

J
Jäkel, Thomas 57, 95
Judmaier, Peter 15

K
Kiesling, Elmar 3
Kietreiber, Tobias 15
Klausner, Lukas Daniel 15

Kovac, Fabian 15
Krau, Tatjana 47
Kromp, Florian 30
Kubesch, Jonas 3
Kumar, Mohit 30

L
Ladner, Sara 15
Lammer, Gregor 30
Lischka, Fabio 79
Litschka, Michael 15
Lüdemann-Ravit, Bernd 47

M
Micheler, Matthias 87
Müller, Benedikt 95

P
Priebe, Torsten 15

R
Rehrl, Jakob 67
Ricken, Tobias 47

S
Schäfer, Georg 67
Schirmeier, Frank 57, 79, 95
Schwarz, Andreas 79
Seliger, Raphael 67, 87
Serles, Umutcan 3
Slijepčević, Djordje 15

T
Toma, Ioan 3

U
Unsin, Sebastian 95

V
Valentinitsch, Alexander 30

W
Waclawek, Hannes 110
Waheed, Faiza 73

Wald, Christoph 79
Wiesner, Dominik 79

Z
Zeppelzauer, Matthias 15

GPSR Compliance
The European Union's (EU) General Product Safety Regulation (GPSR) is a set
of rules that requires consumer products to be safe and our obligations to
ensure this.

If you have any concerns about our products, you can contact us on

ProductSafety@springernature.com

In case Publisher is established outside the EU, the EU authorized
representative is:

Springer Nature Customer Service Center GmbH
Europaplatz 3
69115 Heidelberg, Germany

www.ingramcontent.com/pod-product-compliance
Lightning Source LLC
Chambersburg PA
CBHW071816090925

32286CB00029B/885